Style and Readability in Technical Writing

A Sentence-Combining Approach

Style and Readability in Technical Writing

A Sentence-Combining Approach

James DeGeorge
Indiana University of Pennsylvania

Gary A. Olson
University of North Carolina at Wilmington

Richard Ray
Indiana University of Pennsylvania

McGRAW-HILL, INC.
New York   St. Louis   San Francisco   Auckland   Bogotá   Caracas
Lisbon   London   Madrid   Mexico   Milan   Montreal   New Delhi
Paris   San Juan   Singapore   Sydney   Tokyo   Toronto

STYLE AND READABILITY IN TECHNICAL WRITING
First Edition
987654

Copyright © 1984 by McGraw-Hill, Inc. All rights reserved.
Printed in the United States of America. Except as permitted
under the United States Copyright Act of 1976, no part of this
publication may be reproduced or distributed in any form or by
any means, or stored in a data base or retrieval system, without
the prior written permission of the publisher.

Library of Congress Cataloging in Publication Data

DeGeorge, James 1936-
  Style and readability in technical writing.
  1. Technical writing.   1. Olson, Gary A., 1954-
II. Ray, Richard, 1933-   III. Title
T11.D37   1983   808'.0666   83-11068
ISBN 0-07-554380-X

Preface for the Instructor

This book is based on an idea that is both new and old, innovative yet fundamental. Sentence combining as a systematic way to teach composition is relatively new. Yet the real basis for sentence combining and for this book is probably the most fundamental technique used by technical writers, or any writers interested in precise communication.

That technique is *tinkering*. Writers tinker with their prose when they examine and listen to the structures they have created and consider stylistic options based on other, often more economical, clear, or "readable" ways of delivering their message. In a sense, they combine and recombine stylistic options that have become available to them through their experience with language and their innate ear for it.

A sentence-combining program, or text, provides those who are learning to write with the opportunity to experience and learn systematically from tinkering according to a designed, incremental sequence and under the watchful eye and ear of an instructor experienced with language.

For those of you who haven't used sentence combining, a word of background. Research indicates that sentence combining works at least as well as, and in most cases better than, the more conventional methods for teaching writing at the sentence level.[1] We believe that sentence combining works because it taps the innate ability of language users to tinker with expression until they get it right.

One of the advantages of this trial-and-error approach is that it minimizes the need for abstract grammatical knowledge and terminol-

---

1. Reviews of this research and experimentation can be found in a number of places. We'll cite two: M. F. Stewart, "Sentence Combining: Its Past, Present, and Future," *English Quarterly*, Winter 1979, pp. 21–36; and Andrew Kerek et al., "Sentence Combining and College Composition," *Perception and Motor Skills*, December 1980, pp. 1059–1157.

ogy. Since students are tinkering with language on the basis of innate knowledge, they can learn and execute various combining operations intuitively, even if they haven't mastered the grammatical explanations underlying them. For that reason, it is possible to keep grammatical terminology to a minimum, as we have tried to do in this book.

Rationale for This Book

Many teachers of technical writing find that students with advanced technical knowledge often have difficulty composing clear, readable, technically complex sentences. But traditional technical-writing texts don't devote much space to this problem because they emphasize form and must assume a certain amount of stylistic sophistication on the part of students. *Style and Readability in Technical Writing* fills the gap left by standard technical-writing texts, deals squarely with the problem of students whose technical knowledge has outstripped their ability to write professionally adequate technical sentences and paragraphs, and serves as a supplement to most technical-writing texts now on the market. It does this by combining technical subject matter with an instructional program of sentence combining. No other book takes this approach.

A further rationale for this book is that, unlike other sentence-combining texts, *Style and Readability in Technical Writing* emphasizes and is designed to produce the pragmatic writing virtues that technical writers consider paramount: clarity, economy, and readability.

Style and Readability in Technical Writing brings together current research in stylistics and readability. It does this by identifying and teaching the sentence structures that linguists, reading experts, and cognitive psychologists have found to be relatively more readable than others. After introducing students to the basic operations of sentence combining, the book presents guidelines and exercises for writing highly readable sentences. It then goes one step further toward teaching readability by illustrating ways to build coherence into paragraphs.

We have designed this book for students who must write about complicated subjects in the most readable, economical way possible and who must develop the ability to write this way easily and routinely. The book is also designed for the technical-writing teacher, who must concentrate on form as a primary objective. Busy technical-writing teachers often feel they don't have time to teach sentences or sentence editing. This book will give instructors who face this constraint a supplementary tool.

Format of the Book

The book begins with an "Introduction for the Student," which discusses the qualities of effective writing and the general mechanics of sentence combining. Its purpose is to prepare the student for the first chapters of the book.

Units 1-7 contain instructions and exercises in a variety of general sentence-combining operations and strategies. Unit 1 deals with addition/deletion operations, and Units 2-5 are devoted to specific addition/deletion strategies. Unit 6 focuses on two embedding strategies. Unit 7 covers transposing strategies.

Units 8-10 supplement the earlier chapters by presenting additional ways to achieve readability in sentences and paragraphs. Finally, Unit 11 offers many sentence-combining exercises in typical technical writing formats, such as abstracts, transmittal letters, and executive summaries. This unit provides the student with ample opportunity to practice the techniques taught in the earlier sections of the book. A glossary of terms appears at the end of the text.

The Exercises

At the end of Units 1-7 are Trial Run Exercises: sets of short sentences to be combined into one sentence according to the operation or strategy in question. Space is provided in the text for students to perform these exercises. Written in the text, the completed exercises will provide an accessible review and refresher for students, as well as a convenient format for you to check quickly students' competence in the given combining tactic, before students proceed to the longer Combining in Context exercises.

Units 1-7 conclude with Combining in Context exercises, which are to be done outside the text. They provide an opportunity for students to use all operations and strategies encountered up to that point in the text. We hope that the Combining in Context exercises will provide interesting technical passages with a beginning, a middle, and an end, and a chance for students to use their technical knowledge to help solve the syntactic puzzle that each set of short sentences possesses.

Units 8-10 also include exercises that enable students to practice the strategies learned in the text, and Unit 11 consists solely of exercises.

A Plan for Using This Book

As an aid to technical-writing teachers, who have more to cover than sentence effectiveness, economy, and readability, we suggest three different ways to use this book: You can (1) follow the plan of *Style and Readability in Technical Writing* as laid out in the table of contents; (2) begin with Unit 8, "Readability Guidelines"; (3) use the units in this book according to the sequence dictated by your primary text. Here are a few detailed suggestions.

Following the Sequence Presented

First, if you follow the sequence built into *Style and Readability in Technical Writing*, two whole meetings and small segments of three additional meetings should be sufficient to cover Units 1-5, which give students the base needed to understand all the sentence-combining strategies covered in the text. When possible, class instruction should include oral sentence combining so that students can develop an ear for the combinations they are putting together. You can then cover the remaining units with a minimal amount of class time devoted to teaching the text.

The book is designed to be self-instructional after you have introduced the concept in any given unit. At that point the exercises will take over and lead the student incrementally through the strategy involved. You can, of course, intervene at any point, have students perform exercises in class (in a ten-minute segment, say), make suggestions, and encourage students to explore other sentence combinations that will achieve the intended objectives.

Indeed, it is important for you to monitor students' progress by checking their exercises. As noted, in Units 1-7 we have provided Trial Run Exercises to make it easy for you to check quickly to determine whether or not students have mastered the strategy in question. Grades need not be given on individual exercises, but an overall completion score is not a bad idea. If students can complete the exercises with about 85 percent accuracy, then they probably have a good grasp of the material.

A little before mid-term, some class time should be spent on Unit 8, "Readability Guidelines." By this time some papers will have been written and returned, and students will be more aware of the usefulness of the guidelines and advice presented there. Units 8-10 also offer an opportunity for individualized instruction in those cases where you have identified individual writing problems in the papers done early in the term. Editing and rewriting can be keyed to the appropriate sections or concepts covered in these units.

Beginning with Unit 8, "Readability Guidelines"

Those of you who elect to begin with Unit 8, "Readability Guidelines," can use the guidelines as sentence-editing tools. The guidelines allow you to begin directly with systematic instructions for correcting or revising poorly written sentences in assignments, since they discuss matters such as strong active verbs, delayed subjects, noun clusters, nominalizations, jargon, sentence length, and so on. Then you can either teach the sentence-combining strategies at the beginning of the book after covering the guidelines or intersperse the sentence-combining units between the guidelines. In this way you deal immediately with sentence problems, then introduce students to sentence structure options that are not yet part of their repertoires.

Following that, you can turn to Units 9 and 10 to cover nominalizations, paragraph development and cohesion, transitions, and linkage of "old" and "new" information in larger discourse blocks.

Following Your Primary Text

The third plan for using this book is to fit its units into the plan determined by your primary text.

The Combining in Context exercises and the exercises in Unit 11, when completed, produce an example of one of the standard formats taught in technical writing—for instance, abstracts, transmittal letters, and executive summaries. One way to use this book in conjunction with other texts is to assign students one of the sentence-combining exercises cast in a particular format at the same time that you cover that format in the primary text. Students can then critique the example produced from the sentence-combining exercise according to the instructions given in your coverage of that format. Thus, at the same time that students are learning a technical-writing format, they can be learning the sentence-combining strategy that is involved in the production of that format. At the end of this preface, we provide two indexes of exercises according to format classification.

All technical-writing texts have a chapter or two on clarity, correctness, diction, sentence structure, and so on. To help you match the topics in these sentence-editing chapters and supplement them, we also provide a cross-referenced list of topic headings found in most technical-writing texts and the units in *Style and Readability in Technical Writing* that deal with those topics.

Perhaps we should end this discussion with a word about the syntactic puzzle factor in sentence combining. Our experience, and that of other sentence-combining enthusiasts, tells us that students are fascinated with the syntactic-puzzle approach to composing (call it a Rubiks Cube effect, if you will). Each sentence-combining set, like any puzzle, offers a hard-to-resist challenge and invites students to tinker with the pieces of the puzzle until they reach a solution. Fortunately for us all, the tinkering is exactly what helps students develop syntactic skill. And we think that the built-in puzzle will especially appeal to technical-writing students who plan to make a profession out of a particular kind of puzzle solving.

We would like to thank the following people for encouraging the project in its early stages: Bertie Fearing, East Carolina University; Richard P. Profozich, Prince George's Community College; and William E. Smith, Utah State University. For reading the manuscript and providing detailed comments, we also thank Sally Brett, East Carolina University; Hugh Burns, United States Air Force Academy; Debra Journet, Clemson University; and Julie Lepick Kling, Texas A&M University.

We would be remiss if we failed to acknowledge our gratitude to the many writers and scholars who in some way influenced our work. Among those in the technical-writing field are Donald Cunningham, David Fear, Bertie Fearing, Kenneth Houp, John Lannon, Ann Laster, J. C. Mathes, Gordon Mills, John McNair, Steven Pauley, Thomas Pearsall, Nell Ann Pickett, Jack Selzer, Bill Smith, W. Keats Sparrow, and John Walter. In addition, we are especially indebted to the work of Noam Chomsky, Francis Christensen, Donald Daiker, František Daneš, Andrew Kerek, George Klare, Max Morenberg, Frank O'Hare, Bill Strong, and Joseph Williams. In no way do we imply that these scholars endorse this text; rather, their contributions to scholarship and to our understanding of the writing process have proved invaluable.

Also, Rita Fortune deserves special recognition for her excellent secretarial assistance, as do Steven Pensinger, Elisa Turner, and Laurel Miller, our editors, for their enthusiastic encouragement.

Finally, we wish to thank Frankie DeGeorge, Marlyne Olson, and Marlene Ray, whose collective patience and support made a tedious task less arduous.

James DeGeorge
Gary A. Olson
Richard Ray

Sentence-Combining Exercises Listed by Format

Format	Page(s)
Description	18, 22–23, 29–30, 30–31, 41–42, 60, 93–94, 94–97, 174–179, 179–182
Definition	23–24, 31–32, 32–34, 58–60, 75–77, 77–79
Correspondence	
Letter of Transmittal	155–156, 157–158, 158–159
Letter of Application	165–167, 167–169
Abstract	152–153, 153–154, 154–155
Technical Report	
Task Statement	19
Summary	19–21, 21–22, 159–160, 161–162, 162–165
Conclusion	170–171, 171, 171–173
Narration	39–41, 42–43, 55–57, 57–58, 73–74, 74–75, 79–81, 91–93, 97–99, 99–100

Sentence-Combining Exercises
Listed by Title and Format

Exercise	Format	Page(s)
"The Pressurized Radiator System"	Description	18
"Report on the Model 12-A Cider Press: Statement of the Problem"	Task Statement	19
"Summary of Results: Cutting and Pressing Test of the Model 12-A Cider Press"	Summary	19–21
"Summary of Results: Start-Up Cost Negotiations with White Swan Enterprises"	Summary	21–22
"Specialty Steels"	Description	22–23
"Extended Definition: Niobium"	Extended Definition	23–24
"The New Fuels"	Description	29–30
"Consolidating Concrete"	Process Description	30–31
"Quality Circles"	Definition	31–32
"Extended Definition: The Daisy Wheel"	Extended Definition	32–34
"Cellular Radio"	Narration	39–41
"Automatic Hemming Machine"	Description	41–42
"Retort Pouches"	Narration	42–43
"Adaptive Use of Historic Railroad Stations"	Narration	55–57
"Ann Arbor Circulation Plan for Future Traffic"	Narration	57–58
"Arc Welding"	Process Definition	58–60
"Calibration and Standardization"	Process Description	60
"Government Efforts to Control Health-Care Costs"	Narration	73–74
"Technology and Health Costs"	Narration	74–75
"The Computer Tomography (CT) Scanner"	Definition	75–77

Exercise	Format	Page(s)
"Forging and Hot-Metal Stamping"	Process Definition	77–79
"An Overview of National Transportation Expenditures"	Narration	79–81
"Maritime Cargo Control"	Narration	91–93
"Handling of Samples"	Process Description	93–94
"Two Kinds of Handsaws"	Description	94–97
"Transportation and the Elderly"	Narration	97–99
"Transportation and Air Pollution"	Narration	99–100
"The GA-40 Flashlight"	Description	174–179
"Introduction to the Drafting Compass"	Description	179–182

Index for Cross Referring

Topics and Subtopics Often Found in Technical Writing Texts	*Relevant Units in* Style and Readability in Technical Writing
Clarity	
Sentence Level	Unit 8 (Guidelines 1–10)
Coherence	
Paragraph Level	Unit 10
Sentence Level	Units 1–7
Conciseness	
Sentence Level	Units 1–7
Word Level	Unit 8 (Guidelines 1, 3, 4, 6, 9, 10), Unit 9
Readability Indexes	
Gunning, and so on	Unit 8 (Guidelines 1–10)
Sentences	
Length	Introduction, Unit 8 (Guidelines 9, 10)
Order	Units 1, 2, 3, 4, 5, 7
Independent Clause	Units 1, 5
Dependent Clause	Units 1, 2, 4, 5, 8 (Guidelines 6, 7, 10)
Modifiers	Units 1, 2, 3, 4, 7, 8 (Guidelines 6, 7, 9, 10)
Parallel Structure	Units 1, 4, 5, 7
Tone	Unit 8 (Guidelines 3, 9, 10)
Words	
Active/Passive	Unit 8 (Guidelines 1, 2)
Diction, Jargon	Unit 8 (Guidelines 3, 4)

Contents

Preface for the Instructor	v
Introduction for the Student	1
Unit 1 **Combining with Addition/Deletion**	5
Unit 2 **Combining with the Wh-Connection**	13
Unit 3 **Combining with the Noun/Noun Connection**	25
Unit 4 **Combining with the Ing/Ed-Connection**	35
Unit 5 **Combining with Miscellaneous Addition/Deletion Strategies**	45
Unit 6 **Embedding Clauses and Phrases**	61
Unit 7 **Transposing Strategies**	83
Unit 8 **Readability Guidelines**	101
Unit 9 **Nominalizations**	121
Unit 10 **Paragraph Cohesion**	135
Unit 11 **Open Exercises**	151
Glossary	183

Style and Readability in Technical Writing

A Sentence-Combining Approach

Introduction for the Student

The purpose of *Style and Readability in Technical Writing: A Sentence-Combining Approach* is to help you write more effectively through sentence combining.

Although you can find many explanations of what writing effectively means, we prefer to define it in terms of the reader. Ultimately, the reader decides whether your writing is clear, whether you have presented the information or ideas in a style that is satisfactory, and whether you have a sound understanding of your subject. Once a piece of writing leaves your desk, it must stand on its own, and the judge of its effectiveness will be the reader.

When writing on the job you sometimes will have the good fortune to receive feedback about your writing. At the time this may seem to be ill fortune, since ineffective writing tends to produce feedback more quickly than effective writing does. Often, silence about your writing efforts will pass as praise or, at least, as satisfaction. After all, those who will read your writing have many duties and will not, as a rule, spend large amounts of time thinking specifically about your writing style. Much of the time, their primary interest will be in your ability to convey information clearly and accurately.

Unfortunately, an ineffective style calls attention to itself. It creates grumbling in the mildest-mannered reader about such matters as choppy sentences and awkward phrasing. If this is your reader's reaction, clarity and accuracy may be there, but the "noise" created by your style interferes and makes the reader work hard to find your meaning.

Being able to gauge a reader or an entire audience and then to write in a manner that both informs and satisfies (or, better yet, pleases) them requires experience. A good starting point is to become sensitive early in your training to the presence of the reader and to learn as much as you can about the problems involved in producing readable writing. Some sections of this text have been devoted to that topic.

With a little effort, you can put into your writing arsenal a variety of strategies that can be used to generate stylistic options as you write. Knowing how to use these strategies will build your confidence as a writer. By providing stylistic options, these strategies also will give you more flexibility in handling the problems that arise as you write. One way to learn strategies is through practice in sentence combining.

Sentence combining is not some sort of writers' miracle that will answer all the questions you may have about writing. Neither will it resolve all the problems you may encounter in any given piece of writing. Yet the potential applications of sentence combining are quite impressive. You must learn only a few basic combining techniques, and these will lead you to almost limitless stylistic options for structuring sentences.

There is nothing complex about sentence combining. To demonstrate its simplicity, consider this group of short sentences:

> Packages arrive at the company warehouse.
>
> More than a thousand packages arrive every day.
>
> Some of the packages are as small as watch boxes.
>
> Some of the packages are the size of kitchen stoves.

If you combine these into a single sentence, you might come up with the following possibilities:

> Every day more than a thousand packages, some as small as watch boxes, some as large as kitchen stoves, arrive at the company warehouse.
>
> More than a thousand packages, some as small as watch boxes, others as large as kitchen stoves, arrive at the company warehouse every day.
>
> More than a thousand packages, some as small as watch boxes, some as large as kitchen stoves, arrive at the company warehouse every day.
>
> At the company warehouse more than a thousand packages, some as large as kitchen stoves, some as small as watch boxes, arrive every day.
>
> Among the thousand or so packages that arrive every day at the company warehouse are those so small they could contain nothing

larger than watches, while others are big enough to hold kitchen stoves.

You may even have other ways of combining these sentences. These five sample combinations do not exhaust the options available to you, but rather, they show that there are many ways of conveying the information contained in the four original sentences.

In these five combined sentences, the basic rhetorical strategies discussed in this book—deletion, addition, embedding, and transposition—have been used. It should not surprise you to learn that in the past you have employed these strategies in your writing and revising. Tinkering with sentences always brings one or another—or all—of these strategies into use, even if you are unaware that you are using strategies, let alone that they have names.

After you have worked through the early chapters of this book, your use of the strategies will be more conscious and have more direction. Yet the strategies will remain easy to use, and you will find yourself applying them more often and more productively in your own writing.

Despite what you will learn here, a basic truth about all kinds of writing also applies to sentence-combining strategies: Nothing anyone can say about writing will be true all the time. You will, of course, want to use what you learn in this book whenever you can, but do not make the mistake of always applying the advice without first considering whether it works in a particular sentence.

For example, sentence combining is a process designed to produce longer sentences. The longer sentences that result from sentence combining are often more effective and, more important, more readable than groups of shorter sentences containing the same information. However, experienced writers know that long sentences are not necessarily more effective sentences. And they also know that short sentences are not necessarily effective sentences. Both a long and a short sentence can be unclear or hard to read. Effective writing always contains a mixture of sentence lengths.

Sentence length is never as important as sentence economy or economy of language. Often, one long sentence can say the same thing in fewer words than a group of short sentences can. Sometimes, long sentences can be reduced by recombining the short in them to eliminate redundancy. The important thing to keep in mind is that you can often reduce the number of words by combining and squeezing meaning into more economical sentence structures. But there are limits. Some sentences are more readable than others, and sentence combining can get out of control.

Knowing how long or how short a sentence should be is a matter of judgment. If you have not made a practice of doing so, begin now to examine what you read each day. Notice the number of short sentences. How often are they used? When does the writer use them? Does the writer attempt to blend short and long sentences to achieve some sort of stylistic balance? Do the shorter sentences seem more effective because of the longer sentences?

If you carefully examine what you read, you will quickly learn something about variety in sentence length. You should examine, in particular, as much writing as you can from your own field. Then you will be able to apply intelligently one of the important rules of sentence combining: *Sometimes the proper stylistic strategy is not to combine at all.*

Experience that comes from intelligent observation and practice will tell you when to exercise this option. And when you are not exercising this option—which will be much of the time—skill in sentence combining will provide you with the many stylistic options necessary for producing effective writing.

Now, on your own, see how many different ways you can combine the following short sentences into a single, longer sentence:

Each package has a code.

The code is printed on one side of the package.

The code is printed in numbers.

The numbers are large.

The numbers are black.

The numbers tell what is in the package.

After you have produced several longer sentences from these short ones, decide which of your creations you like best and be ready to explain why.

A good way to test sentences for effective phrasing is to read them aloud. If you have not done this before, doing so may seem strange or funny at first, but these reactions will pass quickly. Eventually your ear will become helpful in forming judgments about the phrasing of sentences. As you continue to do this, you will be able gradually to lower your voice to a whisper and still hear the sentence critically. Eventually you will be able to do the same thing by reading silently, allowing your ear to hear the rhythms and patterns of the sentence. This last will take time to achieve, but it will come. In the meantime, remember that it is important to hear how a sentence goes together.

Unit One

Combining with Addition/Deletion

In English, all sentences, spoken or written, are combined by a number of fundamental operations. As you encounter some of these operations in the next few pages, you'll see that they are not really new to you. To some degree, you have been using them all your life, and you probably do them automatically. The fact that these operations are programmed into the language center of your brain means that you will not have to learn new or difficult rules of grammar. Instead, you will be activating and bringing to a higher performance level the competence for language that you already have. The optimism that underlies this statement, and indeed, this entire book is based on the assumption that the operations you are about to encounter are already part of your language competence. In addition, the systematic exercises, which build on each other, will help you see that you can create more efficient, readable, and sophisticated sentences than you might have thought yourself capable of composing.

We won't cover all the possible combinations of operations. There's no need for a complete listing, and you certainly don't have to memorize the definitions or names of the operations. To learn to execute the operations that are discussed here, you need only conscientiously perform the exercises in each unit. You will soon get the hang of tinkering with the short sentences that you are asked to combine. Tinkering with the sentence problems is important because they're the core of this book. Your built-in knowledge of language will allow you to proceed largely by instinct, however, and your ear, as well as your instructor, will tell you when the combination you have produced is an effective sentence.

There are no "right" and "wrong" answers to the exercises in this book. Rather, you will want to strive for economy, clarity, and accuracy. If a sentence does not sound as though it has these qualities after you or your classmates have constructed it, chances are that the sentence should be recombined.

We have suggested that you listen to your combinations. Most of the exercises in this book should be done orally with your classmates, if possible. Hearing these sentences will sharpen your judgment. You'll be surprised how often class members will agree on which sentences sound better than others, and you'll find that there is often consensus as the collective ear of the class develops.

Sentences, as we have said, are combined by a limited number of fundamental operations, some of which occur in pairs. *Addition/deletion* is the first of these pairs to be considered. To see how these two operations work together, consider this example:

> All glassware should be cleaned and rinsed with acid.
>
> The glassware is nondisposable.
>
> The cleaning and rinsing should be thorough.
>
> The acid should be nitric.
>
> The acid should be a 50 percent solution.

Obviously, these sentences are too short and choppy. Upon reading them, you probably automatically want to combine them. Here is an example of how these sentences might be combined:

> All nondisposable glassware should be thoroughly cleaned and rinsed with a 50 percent nitric acid solution.

This combination is a result of *deleting* (cutting) material from some of the sentences and *adding* material to the main sentence. Let's see these operations graphically:

> All glassware should be cleaned and rinsed with acid.
>
> ~~The glassware is~~ nondisposable.
>
> ~~The cleaning and rinsing should be~~ thorough.
>
> ~~The acid should be~~ nitric.
>
> ~~The acid should be a~~ 50 percent solution.

The slashes through the letters indicate the words that have been deleted. Now consider the additions:

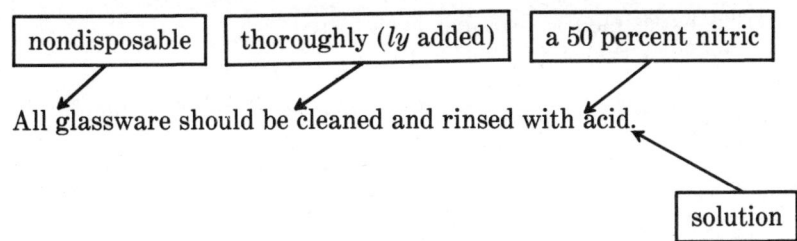

Nondisposable is added as an adjective to *glassware; thoroughly* is added as an adverb to *cleaned* and *rinsed;* and *a 50 percent* and *nitric* are added as adjectives to *acid*, which becomes an adjective modifying the added word *solution*.

Let's see these operations in another example:

The volume serves as a compression chamber.

The volume is left.

The volume is left at the top of the cylinder.

Delete:

~~The volume is~~ left.

~~The volume is left~~ at the top of the cylinder.

Add:

The volume left at the top of the cylinder serves as a compression chamber.

Now that you've studied some examples of addition and deletion, it's time to try a few on your own. Where space has been left, you can do the exercises in the book; longer exercises should be done outside the book. In the following examples, remember to *delete* material from one sentence and *add* to the main sentence:

1. The cast-iron plates must be lifted into place.
   A crane must lift the plates.

2. The experimental drug has been tested.
   The FDA thoroughly tested the drug.

3. Dupont quietly bought the patent.
   The patent was for a new polymer.

4. An accurate miter is essential.
   The miter is at Joint B.

5. Use a pair of tin snips to cut along the outline.
   The outline is scribed on the copper sheet.

6. The brass safety valve will alleviate the problem.
   The problem is one of pressure.

7. Carefully determine the exact bridge span.
   Determine the span before beginning additional calculations.

8. The antenna must rise above all obstructions.
   The obstructions are immediate.

9. Laser technology is rapidly making wire obsolete.
   The wire is copper.

10. The beaker must not be heated.
    The heating is beyond 120° Fahrenheit.

TRIAL RUN EXERCISES

Combine each of the following sets of sentences into one economical, clear sentence using the addition/deletion operation. Try to reduce the total number of words without losing any information.

1. The thermocouple protects homeowners.
   The thermocouple is on a furnace.
   The protection is against explosion.
   The explosion is of gas.

2. An adjustable wrench loosens and tightens.
   The loosening is of nuts and bolts.
   The tightening is of nuts and bolts.
   The nuts and bolts are of many different sizes.

3. Electrical cord is used in lamps.
   The cord is flexible.
   The cord is called zip cord.

4. Zip cord contains two strands of wire.
   The strands are covered with insulation.
   The insulation is plastic.

5. Each strand is made up of wires.
   The wires are many.
   The wires are very fine.

6. Zip cord gets its name from a groove.
   The groove is in the plastic insulation.
   The groove is between the two main strands.

7. Pascal, a programming language, takes its name from Blaise Pascal.
   The language is for computers.
   Blaise Pascal was a Frenchman.
   Blaise Pascal was a mathematician.

8. Everyone's blood carries antibodies.
   The antibodies are in the blood serum.

9. Antibodies protect us.
   The protection is against diseases.
   The protection is by attacking.
   The attacking is of disease antigens.

10. Bubblesort is a computer routine.
    The routine is for sorting.
    The sorting is internal.

11. A fluorescent light is a tube.
    The tube is glass.
    The tube has cathodes.
    The cathodes are at each end.

12. A fluorescent tube uses less energy than an incandescent bulb.
    The energy is to produce the same amount of light.

13. An electric current arcs.
    The arcing is from an electrical contact at one end.
    The arcing is to a contact at the other end.
    The arcing is through a gas.
    The gas is in the fluorescent tube.

14. The electrical arc produces light.
    The arc is in a fluorescent tube.
    The light is ultraviolet.
    The ultraviolet light is invisible.

15. A fluorescent lamp glows.
    The glowing is with light.
    The glowing comes from phosphors.
    The phosphors are a coating on the glass tube.
    The coating is on the inside of the tube.
    The phosphors glow when touched.
    The touching is by ultraviolet light.

16. Modern sonar equipment is sensitive.
    It is sensitive enough to distinguish.
    The distinguishing is among species.
    The species are of fish.

17. Some metals are strategic metals.
    Rhenium is a strategic metal.
    Germanium is a strategic metal.
    Tantalum is a strategic metal.
    So is selenium.

18. Beekeepers use centrifuges.
    The use is for extracting honey.
    The extracting is from the comb.

19. Dipping a pipe galvanizes it.
    The pipe is a steel pipe.
    The dipping is in molten metal.
    The molten metal is zinc.

20. Sweating is the name of a process.
    The process is a soldering process.
    The process is for joining sections.
    The sections are tubing.
    The tubing is copper.

Unit Two

Combining with the Wh-Connection

As you compose sentences, you will find yourself focusing on some facts and assertions while subordinating others. It is only natural to construct sentences so that some parts contain primary information while other parts contain secondary information, which either modifies or amplifies the primary data. One way to establish such a relationship within a sentence is to use a *wh-clause,* a group of words that, like an adjective, modifies a noun or a pronoun. We call this structure the *wh-connection* because four of the five connective words introducing wh-clauses begin with the letters *wh: which, who, whom, whose.* Wh-clauses also begin with *that,* so we include *that* as a signal under the wh-connection.

Here are some examples of sentences containing wh-clauses:

1. An "active network" generally possesses characteristics *that are different* from those of simple RLC circuits.

2. Figure 2 illustrates the end view of an ogee gutter, *which is drawn* to a five-inch scale.

3. Joe Tush is an aerospace engineer *who is destined* to become a leader in solid-fuel technology.

4. Dr. Johnson, *whom you have met,* will direct the project.

5. Please submit all feasibility proposals to the project engineer, *whose office is on the fifth floor.*

Notice that the wh-connection in each example refers back to a noun in the same sentence. Reading from sentence 1 through 5, you will see that *that* refers to *characteristics*, *which* to *gutter*, *who* to *Joe Tush*, *whom* to *Dr. Johnson*, and *whose* to *engineer*. We call wh-clauses *clauses* because they contain both a subject and a predicate. For example, the subject of the wh-clause in sentence 1 is *that*, and the predicate is *are different*. The subject of the clause in sentence 2 is *which*, and the predicate is *is drawn*. And the subject of the clause in sentence 3 is *who*, and the predicate is *is destined*. But wh-clauses are not sentences by themselves because they do not express complete thoughts. They must be connected by the wh-connectors to the primary part of a sentence. For example, you wouldn't write the following statement like this:

Which is a form of energy.

Rather you would write this:

Radiation, *which is a form of energy*, can be beneficial.

The wh-clause must be *added* to the primary part of the sentence.
When you add a wh-clause to a sentence, you may be combining two or more sentences. In example 3, there are two shorter sentences:

Joe Tush is an aerospace engineer.

Joe Tush is destined to become a leader in solid-fuel technology.

Combined they read:

Joe Tush is an aerospace engineer who is destined to become a leader in solid-fuel technology.

Similarly, sentence 4 comprises two shorter sentences:

Dr. Johnson will direct the project.

You have met Dr. Johnson.

Combined:

Dr. Johnson, whom you have met, will direct the project.

As you can see, combining a number of sentences into one clear compact sentence with no loss of information is obviously efficient and economical.

It might help you to know something about the five wh-connectors used to introduce wh-clauses. The pronouns *who* and *whom* refer exclusively to persons. *Whose* is a *possessive* pronoun and refers to persons. For instance, in example 5 *whose* shows that the office *belongs* to the engineer. *Which* refers to a thing, such as *gutter* in example 2, and *that* can refer to persons or things. For example, you can say:

Joe Tush is an aerospace engineer *that* is destined to become a leader in solid-fuel technology.

Finally, it is important that you know how to punctuate sentences containing wh-clauses, since a mere comma or two can significantly alter meaning. You have two alternatives: Separate the wh-clause from the rest of the sentence with commas, or embed the clause into the sentence without commas. When you separate the clause with commas, you are telling the reader that the meaning of the entire sentence does not depend on the clause, that the clause easily could be eliminated without altering the meaning of the sentence. For example, eliminating the clause from example 4 does not alter the principal meaning of the sentence:

Dr. Johnson, whom you have met, will direct the project.

Dr. Johnson will direct the project.

The primary meaning of the sentence is that Dr. Johnson will take charge of a certain project. The fact that you know Dr. Johnson is incidental and does not have any bearing on the fact that he will direct the project. Since the clause contains information that is *nonessential*, you can let the reader know this by separating the clause from the main part of the sentence with commas.

Conversely, if the clause contains information that alters the meaning of the sentence once it is omitted, you must *not* separate the clause with commas; omit the commas so that the reader can recognize that the information contained in the clause is *essential* to the meaning of the sentence. In the sentence that follows, you need the wh-connections in order to distinguish between two different Dr. Johnsons. Thus, the two wh-clauses are essential to the meaning of the sentence and are not set off by commas:

The Dr. Johnson who resigned from NASA is not the Dr. Johnson who is directing this project.

Can you see the difference in meaning between the two sentences that follow?

Please submit all feasibility proposals to the project engineer, whose office is on the fifth floor.

Please submit all feasibility proposals to the project engineer whose office is on the fifth floor.

The first sentence tells you that there is one engineer, and his office happens to be on the fifth floor. The second sentence implies that there is more than one engineer but that the particular one you want is the one with his office on the fifth floor. As you can see, the commas are signals, telling the reader whether or not a clause is essential or nonessential to the sentence's meaning.

Remember: A wh-clause contains a subject and predicate, and it is introduced by *that, which, who, whom,* or *whose*. Separate a nonessential clause from a sentence with commas, but do not separate a clause that is essential to the sentence's meaning.

TRIAL RUN EXERCISES

Combine the following sentences using wh-connections. Punctuate each sentence carefully, according to its precise meaning.

1. The plate dissipation is suppressed during receiver periods by a fixed bias.
   The bias is switched on by a relay.

2. Particles can accumulate in the oil.
   The oil should be changed often if you operate the generator in a dusty location.

3. The miner must ensure that no coal falls into the drive assembly.
   A miner is responsible for the conveyor belt.

4. The proposal is directed to a manager.
   The manager must reply within one week of the deadline.

5. The supervisor submits a daily shift report to the floor manager.
   The supervisor's job is to oversee all production personnel.

6. The earliest analog computers used amplifier blocks.
   Amplifier blocks were known as operational amplifiers.

7. The exhaust pipe leads to a filtering system.
   The exhaust pipe could be made of 16-gauge stainless steel.

8. Victor Voss will give you your instructions.
   Victor Voss is your new foreman.

9. We give a pay bonus to a production worker.
   The worker's output exceeds everyone else's.

10. We are all indebted to Roberto Halli.
    Roberto Halli designed and tested the first model.

COMBINING IN CONTEXT

In the following exercises, combine the sentences into one or more paragraphs, using the wh-connection to produce wh-clauses wherever appropriate. All the sentences in each exercise are on one topic, so it is necessary to work on the sentences in numerical order.

The Pressurized Radiator System

1. The modern automobile engine has a cooling system.
   The cooling system is a pressurized system.
   The modern automobile engine operates at much higher temperatures than do older engines.

2. This pressurized system handles the increased heat.
   This pressurized system handles the heat by raising the boiling point of the liquid.
   The liquid cools the engine.

3. The most commonly used cooling liquid already has a high boiling point.
   The commonly used cooling liquid is ethylene glycol.
   The high boiling point is somewhere around 220 degrees Fahrenheit.

4. The average pressurized cooling system raises this boiling point.
   The average pressurized system is in today's cars.
   The raising is by another 40 or so degrees.

5. This leeway permits higher temperatures.
   This leeway is extra.
   The higher temperatures are combustion temperatures.

6. It also helps accommodate higher revolutions-per-minute.
   Higher revolutions-per-minute create much heat.
   But higher revolutions-per-minute enable engines to generate more power.
   The engines are smaller engines.

Report on the Model 12-A Cider Press: Statement of the Problem

1. This report deals with two assignments.
   Our Products and Services staff asked us to handle the two assignments.

2. Both concern the Model 12-A cider press.
   The Model 12-A cider press is built for our company.
   The building is done by White Swan Enterprises.

3. The Model 12-A is a small press.
   The Model 12-A is a hand-operated press.
   The Model 12-A is designed for home use.

4. The first assignment asked us to determine something.
   The first assignment asked how efficiently the Model 12-A press could process apples.
   The apples were cut into chunks instead of shredded.

5. The 12-A has always used an attached shredder.
   The attached shredder prepared the apples for pressing.

6. In addition, our Research and Planning Department developed a cutting mechanism.
   They developed the cutting mechanism recently.
   The cutting mechanism is less expensive to manufacture.

7. The second assignment authorized us.
   The authorization was to negotiate a tentative agreement.
   The agreement was on start-up costs.
   The agreement was with President Walters of White Swan Enterprises.
   President Walters had raised questions about the cost.
   The cost was for converting the Model 12-A from shredding to cutting.

Summary of Results: Cutting and Pressing Test of the Model 12-A Cider Press

1. We pressed twenty one-bushel samples of apples.
   The pressing was done during the week of October 23, 1983.
   Ten of the one-bushel samples were prepared with the new cutting mechanism.

The new cutting mechanism was developed by our Research and Planning Department.

2. The other ten batches were run through the shredder.
The shredder is now used on all Model 12-A's.

3. We carefully controlled the weight per batch used for the tests.
We carefully controlled the varieties of apples used for the tests.

4. Seven of each ten batches were mixtures.
The mixtures were of apples.
The apples are commonly available.
The apples are commonly available in the Northeast.

5. Three of each ten batches contained a single apple variety.
The single variety was the Northern Spy.
The Northern Spy is one of a few apple types.
The few apple types are capable of producing high-quality cider.
The few apple types produce high-quality cider without the addition of other apple types.

6. Both the weight controls and the apple mixtures are discussed.
They are discussed in detail.
The are discussed in the body of this report.

7. The batches consistently produced cider volume.
We prepared the batches with the new cutter.
The cider volume was 86 percent to 91 percent of the volume attained.
The volume was attained from comparable batches.
The comparable batches were run through the shredder.

8. This was somewhat higher than we had anticipated.
We had anticipated 75 percent to 80 percent.

9. The cutting mechanism works considerably faster than the shredder.
This confirms a prediction.
The prediction was made by Research and Planning.

10. The average batch cutting time was four minutes.
This was about two minutes faster than the shredder.
The batch cutting time was with one person feeding the cutter.
The batch cutting time was with one person turning the cutter handle.

11. The cutter worked during the processing.
    The cutter worked without clogging at any time.
    The processing was of the ten batches.

12. The shredder clogged on four occasions.
    This is close to the average.
    The average is for that unit in past tests.

Summary of Results: Start-Up Cost Negotiations with White Swan Enterprises

1. We reached an agreement with President Walters.
   The agreement is tentative.
   The agreement assigns 80 percent of the start-up costs.
   The assigning is to White Swan.
   The start-up costs are for refitting the Model 12-A.
   The refitting is with cutter mechanisms.

2. The percentage will be reduced gradually.
   The reducing will be to 15 percent.
   The reducing will be once we begin to market the Model 12-A with cutters.

3. White Swan will then receive 2 percent of the cost.
   The 2 percent is of each unit sold.
   The cost is the cost to our dealers.
   White Swan will receive the 2 percent until the 15 percent figure is reached.
   Or White Swan will receive the 2 percent for five years.
   One of these will come first.

4. The agreement is subject to certain clauses.
   The clauses relate to sales targets.

5. Our director of finance is now studying this agreement.
   Our legal officer is now studying this agreement.

6. A report was sent to them.
   The report was full.
   The report was sent directly.

7. Attachment A summarizes the agreement.
   Attachment A is at the end of this report.

8. This summary is for the benefit of some people.
   Those people coordinate our marketing efforts.
   The coordinating is with White Swan's assembly services.

9. However, the agreement is not mentioned in the body of this report.
   This report deals exclusively with the details.
   The details are of the Model 12-A cutting and pressing tests.

Specialty Steels

1. Business puts specialty steel to use.
   Industry puts specialty steel to use.
   They put specialty steel to use every day.

2. These two groups need specialty steel.
   They need specialty steel to do things.
   These things could not be done with ordinary steel.

3. Specialty steel must be durable and tough.
   It must be durable and tough to satisfy their needs.
   It must be durable and tough enough to last.
   The lasting is for long periods of time.
   The lasting is under the most extreme conditions.

4. The extreme conditions vary.
   This varying causes specialty steels to come in a variety of forms.

5. This variety allows business to select.
   This variety allows industry to select.
   This selecting is of the kind of specialty steel.
   The specialty steel fits the job.

6. Sometimes, this means choosing a steel.
   The chosen steel is extra hard.

7. Other jobs may demand a steel.
   The demanded steel resists high temperatures.

8. Many times steel is used in an environment.
   The environment contains corrosive or abrasive agents.

9. Specialty steels must replace steels.
   The replacing is in these cases.
   The replacing is of ordinary steels.

10. Ordinary steels would weaken.
    Ordinary steels would wear out.
    The weakening and wearing out would happen quickly.
    The weakening and wearing out would be in such environments.
    Such environments would be harsh.

Extended Definition: Niobium

1. Niobium is a metallic element.
   Niobium is called *columbium* in America.
   This metallic element is used primarily as an alloy.
   The alloying is with steel.

2. Niobium strengthens steel.
   The strengthening is by making steel harder.

3. This added strength means something for products.
   Weight can be trimmed from products.
   The products are made from niobium steel alloys.

4. Niobium also reduces costs.
   The costs are steel-making costs.

5. It hardens without losing strength.
   This means something.
   Niobium steel does not have to be cooled rapidly.
   The cooling is by quenching.
   The quenching is in order to temper the steel.

6. This means savings in time.
   This means savings in labor.

7. Niobium will likely grow more popular.
   The growing will be in the future.
   Niobium is already a popular element.

8. An element will always be in demand.
   The element hardens and lightens steel.
   The demand is in a world.
   The world is faced with problems.
   The problems are energy problems.

9. But niobium is attractive for other reasons.
   The other reasons are in addition or as well.

10. Niobium steel alloys resist heat.
    They resist corrosion.
    They resist these better than ordinary steels.

11. This unusual metallic element is widely available.
    Its availability is to the steelmaking industry.
    The steelmaking industry draws on ore deposits.
    The deposits are large deposits.
    The deposits are in Canada and South America.

Unit Three

Combining with the Noun/Noun Connection

Another method of achieving a compact, economical style in professional writing is to use, when applicable, a noun that renames another noun in the sentence. We call this strategy the *noun/noun connection* because it allows you to add qualifying detail concisely to the main sentence by attaching a noun or descriptive cluster of information to a noun in the main sentence.

Here are some examples of the noun/noun connection:

1. An iron worker's brake, *a machine* for bending metal of various gauges, must be included in the original budget.
2. We should be able to use a cathode-ray tube—*a small television-screen display*—in both models of the proposed product.
3. The voltage regulator, *a device* in every car, is the principal cause of mobile interference in both CW and SSB transmission.
4. The word processor, *a computerized typewriter*, can make any business more efficient by reducing the secretarial workload.
5. The floor technician, *a specialist* responsible for all practical matters in the power plant, must be prepared for all emergencies, including a massive system failure.

Each of the examples contains a noun/noun connection—a group of words that together rename a noun in the sentence. The noun/noun connection tells the reader something about the noun and often defines it further. Notice, too, that the noun/noun connection is set off from the rest of the sentence by commas or dashes.

In example 1, the noun/noun connection further defines a brake; it is described as "a machine for bending metal of various gauges." In example 2, the noun/noun connection renames the cathode-ray tube; it is "a small television-screen display." In example 5, it adds detail by explaining what the floor technician is—"a specialist responsible for all practical matters in the power plant." In each example, the noun/noun renames another noun in the sentence:

the *brake* is a *machine*

the *cathode ray tube* is a *display*

the *voltage regulator* is a *device*

the *word processor* is a *typewriter*

the *floor technician* is a *specialist*

You have the option of placing a noun/noun connection either at the front, the middle, or the end of a sentence. For example, you can take the five earlier sentences and place the noun/noun connection in alternative positions:

1. A machine for bending metal of various gauges, the iron worker's brake must be included in the original budget.

2. In both models of the proposed product, we should be able to use a cathode-ray tube, a small television-screen display.

3. A device in every car, the voltage regulator is the principal cause of mobile interference in both CW and SSB transmission.

4. In order to reduce the secretarial workload and make the business more efficient, we can use a word processor, a computerized typewriter.

5. A specialist responsible for all practical matters in the power plant, the floor technician must be prepared for all emergencies, including a massive system failure.

Your main criterion in deciding where to place a noun/noun connection in a sentence always should be clarity. Which sentence format conveys the intended message in the clearest, most precise manner? The main lesson you should learn from this book is that you have many linguistic tools that you should learn to use. Try various options before deciding how to write your sentence.

There are two advantages to using a noun/noun strategy. First, it allows you to add information to a sentence without causing the sentence to become bloated and hard to read. That is, a noun/noun connection allows you to add information to a sentence economically. Second, it is stylistically pleasing. There is a certain cadence in a sentence containing a noun/noun connection, perhaps because the reader must pause before and after the connection.

TRIAL RUN EXERCISES

Combine each group of sentences into one sentence containing a noun/noun connection. Experiment with placing this connection in various positions within the sentences:

1. The mining engineer will examine the shaft for safety.
   The engineer is Joe Banks.

2. The proposed project will study DNA in wild geese.
   DNA is the genetic building block.

3. The foreman will notify us of any irregularities.
   The foreman is our representative at the project site.

4. The project scientist is studying the effects of nicotine on the respiratory systems of chimpanzees.
   The scientist is Dr. Joseph U. Smoker.

5. Hydromechanics comprises three fields: hydrostatics, hydrodynamics, and hydraulics.
   Hydromechanics is the study of fluids.

6. The original blueprint calls for a T-joint made of 16-gauge stainless steel.
   A T-joint is a pipe intersected perpendicularly by another pipe.

7. Third National Bank has ordered a new ATM for its downtown branch office.
   An ATM is an automatic teller machine.

8. The electrodynamometer employs the interaction of the magnetic fields of two sets of coils.
   The electrodynamometer is an instrument that measures current, voltage, or power.

9. Our laboratory has just purchased a cathode-ray oscilloscope.
   A cathode-ray oscilloscope is an instrument that depicts periodic changes in an electric quantity on a screen.

10. Tune the receiver to its highest frequency band and prepare the antenna coupler.
    The band is a range between 15 MHZ and 30 MHZ.

COMBINING IN CONTEXT

In these exercises, combine the following groups of sentences into a paragraph or paragraphs, using noun/noun connections where appropriate. You should also use wh-connections where needed. Since many of the sentences take their meaning from preceding sentences, work on the sentences in numerical order.

The New Fuels

1. Most people have heard of gasohol.
   Gasohol is a blend.
   The blend is of gasoline.
   The blend is of alcohol.

2. You might even know that the alcohol comes from farm products.
   The alcohol is in this mixture.

3. But most people are not familiar with some other fuel mixes.
   The fuel mixes were recently created.

4. For example, how many of us know about petrocoal?
   Petrocoal is a combination.
   The combination is of gasoline with alcohol.
   The alcohol is made from coal.
   The alcohol is made from natural gas.

5. Or who could name the ingredients?
   The ingredients are in methacol.
   Methacol is a fuel.
   The fuel is recommended by the National Maritime Union.
   It is recommended for powering ships.

6. You might easily identify one ingredient by its name.
   The ingredient is methanol.

7. Yet, you would not likely guess the other ingredient.
   The ingredient is pulverized coal.

8. You would probably have the same trouble.
   The trouble is with Hydro Fuel.
   Hydro Fuel is a blend.
   The blend was created by United International Research, Inc.

9. You could predict that water is one element.
   The predicting would be from the name.

10. But the other two parts might elude you.
    One of the two parts is unleaded gasoline.
    One of the two parts is alcohol.

11. You would have an easier time, though.
    The easier time would be with Coco-Diesel.
    Coco-Diesel is coconut oil.
    Coco-Diesel is diesel fuel.
    Coco-Diesel is a mixture.

12. In this case, the name is a giveaway.
    It is a dead giveaway.

13. But only a short-order cook would have a chance with buckfry.
    Buckfry is the last fuel on this list.
    Buckfry is a new fuel.

14. Who but a short-order cook would think of combining?
    The combining is of diesel fuel.
    The combining is with cooking oil.
    The cooking oil is recycled cooking oil.

Consolidating Concrete

1. You can't just pour concrete.
   You can't just let it harden.

2. You must first spread the concrete by pushing or pulling it.
   The pushing or pulling is with a shovel.
   Or the pushing and pulling is with a concrete hoe.
   The pushing or pulling is until the concrete fills the form.
   The form is the outer framework.
   The outer framework holds the concrete until it dries.

3. This pushing helps consolidate the concrete.
   This pulling helps consolidate the concrete.

4. Then, you must take further steps.
   The further steps are after spreading.

The further steps are to consolidate.
The consolidating is of the concrete.

5. When consolidating, you are working out and separating.
   The working out is of the air bubbles.
   The separating is of any remaining clusters.
   The clusters are of aggregate.
   The aggregate is gravel.

6. You must be especially sure to consolidate.
   The consolidating is of the concrete.
   The consolidating is where the concrete meets the form.

7. The trapped air bubbles will leave honeycombs.
   The leaving will be if you don't consolidate there.
   The honeycombs will be on the concrete at the sides.
   The concrete will be hardened.
   Honeycombs are sections of little indentations.

8. Do the consolidating with a concrete spade.
   A concrete spade is a shovel-like tool.
   It is a tool with flat, steel blade attached to a handle.
   The blade is about four inches wide.
   The blade is about eighteen inches long.
   The handle is a six-foot handle.

9. Insert the spade and push the handle.
   The inserting is into the concrete next to the form.
   The pushing is back and forth.
   The pushing is to your left and right.

10. Do this all along the form.
    Do this using extra care.
    The extra care is at the corners.

Quality Circles

1. American corporations have borrowed a good idea.
   The borrowing was recently.
   The borrowing was from Japanese industry.
   The good idea is quality circles.

2. These circles meet to discuss matters like production.
   They meet to discuss matters like the quality of finished goods.

The meeting is regularly.
These circles are small groups of employees.
The employees are from different job levels.

3. Such groups provide a forum.
The forum is for work-place feedback.
The feedback comes from two directions.

4. During meetings production workers can air complaints.
During meetings they can make suggestions.
The airing and making are to foremen, supervisors, and managers.
The foremen, supervisors, and managers are circle members.
These circle members oversee production.

5. At the same time, these supervisory group members can explain.
The explaining is of company problems and plans.
At the same time, these supervisory group members can ask.
The asking is for reactions.
The asking is for advice.
The reactions and advice are from production workers.

6. The circles have done much to change attitudes.
The changing was by providing two-way feedback.
The attitudes were antagonistic attitudes.
The attitudes were of two company groups.
The two groups have historically held antagonistic attitudes.
The holding has been toward each other.
The two company groups are blue- and white-collar workers.

Extended Definition: The Daisy Wheel

1. A daisy wheel is a small, flat, circular printing wheel.
A daisy wheel is used in some word processors.
A daisy wheel is used in many of the printing units.
The printing units are of desk-top computers.

2. This printing element resembles a daisy.
It also resembles a wheel without a rim.
It resembles both of these.
A daisy is the flower the wheel was named for.

3. The petals radiate outward from the center.
The petals are on a real daisy.
The center is round and yellow.
The center is of the flower.

Combining with the Noun/Noun Connection · 33

4. Each petal on the daisy wheel has a character.
   The character is attached.
   The attaching is on the outer end.
   A character is a letter, number, punctuation mark, or symbol.

5. Some people think the daisy wheel looks more like an actual wheel.
   The actual wheel is especially like an old-fashioned buggy wheel.
   The buggy wheel is with a hub in the center.
   The buggy wheel is with thin spokes.
   The spokes radiate out from the hub.

6. The daisy wheel, of course, has spokes.
   The spokes are many more spokes.
   The many more spokes are than the old buggy wheels had.

7. And the spokes do not radiate outward.
   The spokes are on the daisy wheel.
   The radiating outward is to join a closed circle.
   A closed circle is a rim.

8. The daisy wheel has only a hub.
   The daisy wheel has only spokes.

9. The print characters are attached to the outer ends.
   The outer ends are tips.
   The outer ends are of the spokes.

10. Each petal or spoke is approximately one and one-half inches long.
    Each petal is on the daisy.
    Or each spoke is on the wheel.

11. The daisy wheel spins at high speed.
    The spinning is as a printer operates.
    The spinning is about .060 of an inch from a printing medium.
    The printing medium is an ink or carbon strip.
    The strip is similar to a typewriter ribbon.

12. The tips of the petals or spokes of the wheel press.
    The pressing is against the medium.
    The pressing prints the desired character.
    The printing is onto the paper.

13. The entire daisy wheel moves.
    The moving is as it prints.

The moving is first to the right.
The moving is until it reaches the margin of the paper.

14. Then it reverses and prints.
The printing is of the next line.
The printing is "backward."
The printing is from right to left.

15. Printing both ways adds to the other characteristics.
The other characteristics are high-speed characteristics.
The characteristics are of the daisy wheel.

16. Speeds of 40 or so characters per second are quite common.
Or speeds of more than 500 words per minute are quite common.

17. The daisy wheel can also produce letter-quality copy.
Letter-quality copy is printed pages of quality.
The quality is comparable to those typed on typewriters.
The typewriters are the best electric ones.

18. Daisy wheels come in various styles.
The various styles are of the two common type sizes.
One of these sizes is pica.
And the other is elite.

19. The larger size will print 10 characters per inch.
The larger size is pica.
Elite will print 12 characters per inch.
Elite is the smaller type size.

20. Daisy wheels can be snapped into a printer.
The snapping is in a matter of seconds.
And daisy wheels can be removed.
The removing is just as easily.

21. This allows those people to change the type style or type size.
Those people use daisy wheels.
They are owners of word processors and desk-top computers.
The changing of type style or size is simply by replacing.
The replacing is of one wheel with another.

22. Wheels are also available.
These wheels have special characters.
The special characters are the various symbols.
The symbols are used by mathematicians, scientists, and linguists.

Unit Four

Combining with the Ing/Ed-Connection

A flexible and functional method of attaining conciseness in professional prose is to use verb forms ending in *ed* or *ing*. These are called *participles*. Ing/ed-connections usually function as adjectives. All you really need to know is that you can convert verbs to *ing* or *ed* forms and add the resulting ing/ed-connection to the main sentence. Here are some examples:

1. *Designed* to prevent injury, the drill press guard is inexpensive and easy to construct.
2. The angle-iron frame, *welded* to the main support studs, is sturdy enough to support the electric motor.
3. Under such tension, the drive belt is likely to snap, *spinning* from the back of the press and *flying* toward workers on nearby presses.
4. *Mounting* the converter, he discovered a short between switches 1 and 3.
5. Next, the technician must reattach the door plate to the transformer, *steadying* the plate with the left hand and *tightening* the main lugs with the right hand.

You will notice that often you have the option of placing the phrase containing the ing/ed-connection either at the front, the middle, or the end of a sentence. Here are examples 1, 2, and 5 with the ing/ed-connection placed in an alternative position:

1. The drill press guard, *designed to prevent injury*, is inexpensive and easy to construct.

2. *Welded to the main support studs*, the angle-iron frame is sturdy enough to support the electric motor.

5. *Steadying the door plate with the left hand* and *tightening the main lugs with the right hand*, the technician must reattach the plate to the transformer.

Your choice about where to place an ing/ed-connection in a sentence should depend on which position better enables you to convey your precise meaning clearly and effectively. In example 2, you might decide that it is best to avoid separating the subject from the verb by placing the ing/ed-phrase in front of the sentence:

2. *Welded to the main support studs*, the angle-iron frame is sturdy enough to support the electric motor.

Even though you can often move ing/ed-connections around in a sentence, there are limits. Put in the wrong place, an ing/ed-connection can create ambiguous, unclear, or silly sentences, such as the following:

Fermenting in copper barrels, the technician ran a check on the ethanol content of the mash.

Look how meaning changes with different placement of the ed-connection in the following:

Filed for later reference, the papers were added to the case records.

The papers were added to the records filed for later reference.

In the following sentence, the meaning gets lost because there is nothing in the sentence for the ing-connection meaningfully to connect with:

Adding an extra memory card, the microprocessor was considered adequate.

Using an ing/ed-phrase often allows you to introduce additional information into one sentence rather than using two or more sentences. For example, rather than writing:

The crystal-controlled oscillator is mounted toward the back of the cabinet. It will provide good frequency stability.

You can write:

The crystal-controlled oscillator, *mounted toward the back of the cabinet,* will provide good frequency stability.

Or:

Mounted toward the back of the cabinet, the crystal-controlled oscillator will provide good frequency stability.

Both sentences employing the ing/ed-phrase introduce the same ideas as their two-sentence parent, but they do so economically and clearly. This is a good example of how ing/ed-phrases can add information to your sentences. In some cases using an ing/ed-phrase can help you illustrate most precisely the relationships between the various topics.

TRIAL RUN EXERCISES

In the following exercises combine the short sentences, using ing/ed-phrases when possible. Try placing the same ing/ed-phrase in different positions. Does any one placement seem more effective than others? Why?

1. The cost is judged from formal construction estimates.
   This project will cost no less than $95 million.

2. The elevator is supported at its base by reinforced concrete.
   The temporary elevator will facilitate the transportation of construction materials to the highest floors.

3. The crane is anchored to the building frame.
   The kangaroo crane will assist workers above the tenth floor.

4. First, we sterilized the beaker.
   The sterilization destroyed all harmful organisms.

5. The receiver's oscillator is connected directly to the mixer.
   The oscillator generates a signal of constant amplitude.

6. The assembly is constructed of heavy-gauge stainless steel.
   The retainer and bolt assembly should resist corrosion even from the most concentrated levels of pollutants.

7. The unit will not cause overloads on the main power line.
   The unit is supplied by a secondary power source.

8. Bessemer engineers installed the most sophisticated traffic-light system available.
   The system replaced the antiquated lighting system.

9. We modulated the signal's intensity to 500 Hz.
   We increased the output in regular increments.

Combining with the Ing/Ed-Connection · 39

10. The experimenter chooses the laser's operating frequency.
    The frequency coincides with one of several sodium-line frequencies.
    The experimenter must assure that there is no fluctuation in intensity from the power source.

COMBINING IN CONTEXT

Combine the following sentences into several paragraphs. Use the ing/ed-connection and any other strategy wherever appropriate. Since each sentence takes some of its meaning from previous sentences, be sure to work on the sentences in numerical order.

Cellular Radio

1. Telephones in automobiles are about to become more common.
   Their becoming more common is thanks to a new system.
   The new system is a transmitting system.

2. This new system will use several low-powered transmitters.
   This new system is called cellular radio.
   The use of several low-powered transmitters is to relay telephone calls.

3. These transmitters will eliminate many of the problems.
   The transmitters are strategically located.
   The location is in and around big cities.
   The problems are problems now associated with mobile-phone systems.

4. Present systems rely on a single transmitter.
   The single transmitter is a high-powered transmitter.
   The relying on is to relay phone messages.

5. Unfortunately, the high power can cause problems.
   The high power is of these systems.
   The problems are communication problems.

6. A high-powered transmitter sends radio waves.
   The sending is over long distances.
   The sending over long distances sometimes causes interference.
   The interference is with the signals of other transmitters.
   The other transmitters are a long way off.

7. The big transmitters are assigned just a few radio frequencies.
   The assigning is to avoid this interference.

8. The single transmitter can handle only so many calls.
   The calls are incoming calls.
   The single transmitter is limited to a few frequencies.

9. The transmitter quickly builds up a backlog.
   The building up is during peak hours.
   The backlog creates irritation for callers.

10. For example, a busy executive may arrive at the office.
    His arrival was without completing his call.
    The executive tried to call ahead from his car.

11. Cellular radio handles peak hours.
    The handling is with ease.
    Cellular radio uses a series of transmitters.
    The transmitters are low-powered.

12. Cellular transmissions eliminate interference.
    The interference is with distant transmitters.
    Cellular transmissions have a range of only six to eight miles.
    This range permits the use of a wider band of frequencies.

13. The additional frequencies can handle many more calls.
    The many more calls reduce delays.
    The delays are during peak periods.

14. Cellular radio can also handle calls.
    The calls are simultaneous.
    The calls are on the same frequency.

15. The transmitters will accept calls.
    The transmitters are computer controlled.
    The accepting is of calls only from nearby vehicles.

16. Thus, each call will be picked up.
    Thus, each call will be relayed.
    The picking up is by the transmitter closest to the mobile phone.
    The relaying is by the transmitter closest to the mobile phone.

17. When necessary, the computer passes the call on.
    The passing on is automatically.

The passing on is to the next transmitter.
The passing on is to keep up with the moving car.

18. Cellular radio takes its name.
    The taking is from its system.
    The system is of low-powered transmitters.

19. Each transmitter forms a cell.
    The cell extends for six to eight miles.
    The extending is in every direction.

20. These cells will transform metropolitan areas.
    The transforming will be into large, efficient transmitting systems.
    These systems will make mobile phones more attractive.
    They will be more attractive to business.
    They will be more attractive to industry.
    They will be more attractive to individuals as well.

Automatic Hemming Machine[1]

1. The Hemmer provides precise, quality hemming.
   The Hemmer was discussed in last year's Annual Report.
   The hemming is at speeds faster than manual operations.
   The speeds are three to four times faster.
   The speeds are for a wide range.
   The range is of material thicknesses.
   And the range is in solids, stripes, and plaids.

2. The Reece 64 Hemmer hems top-center shirt fronts.
   It hems underfronts, short sleeves, and shirt pockets.
   It hems by transporting the panels by means of a continuous conveyor.
   The transporting is through a folding device.
   The transporting is into a sewing machine.
   And the transportation is onto an unloader/stacker.

3. The Hemmer can cut the constructions following sewing.
   And the Hemmer can separate the constructions following sewing.
   This eliminates up to two operators in mechanizing handling operations.
   The handling operations are in the production of shirts.
   And the handling operations are in the production of blouses.

[1]Adapted from The Reece Corporation, Annual Report 1981.

4. The Series 64 Hemmer can produce 500-dozen shirt fronts.
   Or the Series 64 Hemmer can produce 1,000-dozen sleeves.
   Or the Series 64 Hemmer can produce 2,000-dozen pockets.
   It can do this on a regular eight-hour shift.
   The producing is while providing a wide range of widths and spacing.
   The widths are folding widths.
   The spacings are needle spacings.

5. Monitoring devices turn off the machine.
   The monitoring devices are automatic.
   The turning off is in the event of thread breakage.
   Or the turning off is in the event of depletion.
   The depletion is of liner supplies.

Retort Pouches

1. Retort pouches have begun to appear.
   Retort pouches are newly developed food containers.
   The appearing has been in supermarkets.
   The supermarkets are located around the country.
   The locating is here and there.

2. These pouches may become a common method.
   These pouches are meant to replace traditional cans.
   The method is for packing vegetables.
   The method is for preserving vegetables.

3. Retort pouches require no refrigeration.
   They are like cans in this respect.

4. However, they do not dent.
   They are unlike cans in this respect.
   The denting is when dropped.

5. They are both tough and flexible.
   They are made of metal foil and plastic.

6. They slip back into shape.
   The slipping is if accidentally dropped.
   The slipping is instead of denting.

7. Their skins make them ideal.
   They are ideal for hikers and campers.

And they are ideal for vacationers.
Their skins are flexible.

8. The pouches can be squeezed into places.
   The pouches are flexible.
   The places are those where a can wouldn't fit.
   The can is rigid.

9. How consumers will accept them is a question.
   The consumers are other consumers.
   The question is another question.

10. Many people reject new products.
    Many people feel more secure with familiar things.
    The rejecting is because new products look strange.
    Or the rejecting is because new products look different.

11. Whether these people will react remains to be seen.
    The reacting is in this way.
    The reacting is to retort pouches.

12. Food processors hope to learn something about these people.
    Food processors have placed retort pouches in a few supermarkets.
    The learning is about what their reactions will be.

Unit Five

Combining with Miscellaneous Addition/Deletion Strategies

In this unit we'll group together a number of addition/deletion strategies that you'll find very helpful to know but that are not used as frequently as the strategies we've already discussed. Although you've performed all these operations at some time in the past, you are probably more familiar with some than others.

Coordinate Strategy

The key in *coordinate strategy* and the signal of this operation is the presence of one of these words: *and, but, for, nor, or*. Let's look at three related but separate sentences:

The same rate law is followed in disorption of electrons.

The same rate law is followed in release of electrons.

The same rate law is followed in concentration of electrons.

Now here are the same sentences combined into one with the use of the coordinate strategy:

The same rate law is followed in disorption, release, *and* concentration of electrons.

You can see that a number of nouns are arranged in a series.

Any part of speech can be coordinated in this way. Here is another example, one that coordinates verbs:

The technician *calibrated* the instruments.

The technician *weighed* the anhydrous HCl.

The technician *prepared* the 10 percent HCl solution.

Combined, it would read:

The technician *calibrated* the instrument, *weighed* the anhydrous HCl, and *prepared* the 10 percent HCl solution.

You'll notice that you can make the coordination strategy more obvious and therefore more effective when you put the coordinate elements—verbs, adjectives, nouns, *ing* words, whatever—into parallel grammatical structures. For instance in the preceding example, all the verbs are in the same tense and end in *ed*. In another case, if you were to make one of the coordinate elements an *ing* word, then you would make all of them *ing* words. Here is another example of parallel structure:

The technician had the responsibility for *calibrating* the instruments, *weighing* the compounds, and *preparing* the solution.

Here is a parallel coordinate strategy using a *to* + verb structure:

The technician had to *calibrate* the instruments, *weigh* the compounds, and *prepare* the solution.

Notice that the initial *to* carries over to each of the three verbs; it is not necessary to repeat the *to* each time.

Parallel coordinate structures can help you avoid awkward sentences such as the following:

Combining with Miscellaneous Addition/Deletion Strategies · 47

The Zappo missile performed with precision, was on target, and better than the engineers expected.

Corrected, this would read:

The Zappo missile performed with precision, landed on target, and operated better than the engineers had expected.

Coordination can also be used to join entire sentences. You are probably already familiar with this. In the past you may have used coordination to combine many sentences in the same way as the following example. Uncombined it reads:

Each segment of rigid cylindrical pipe is assumed to move as a unit.

The motions of joints as well as the internal stresses are computed.

Combined, it reads:

Each segment of rigid cylindrical pipe is assumed to move as a unit, *and* the motions of joints as well as the internal stresses are computed.

Notice the comma after the first sentence and before the *and;* when joining two sentences, a comma must be used with one of the coordinating key words.

In the next example, subjects within sentences are combined with *and*, and entire sentences are combined with *but*. This example combines a deletion and addition strategy:

Stress on the pipe segments is constant.

Strain on the pipe segments is constant.

Forces on the joints or nodes of the pipe are variable.

Delete:

Stress on the pipe segment is constant.

Strain ~~on the pipe segment is constant.~~

Add:

Stress *and* strain on the pipe segment are constant, *but* forces on the joints or nodes of the pipe are variable.

Notice that subjects combined with a coordinate strategy require a plural verb, in this case, *are*.

Subordinate Strategy

We call the next addition operation the *subordinate strategy*. With it you can subordinate a sentence to another by adding connecting words such as:

when	although	though	whereas	where
since	while	as though	if	
because	as	even though	as if	

In the following example, *when* is used to add the first sentence to the main sentence in a time relationship:

Oxidizable organic molecules approach the SnO_2 surface.

The chemisorbed oxygen reacts with them.

The result of subordination:

When oxidizable organic molecules approach the SnO_2 surface, the chemisorbed oxygen reacts with them.

Notice that a comma is used after the *when* clause, which in this case comes before the main clause.

The choice of connector can affect the meaning of the sentence. In the following example, note how the choice of the connector word determines the logical relationship of the addition:

You can reasonably predict the photoconductive effect of ZnO and TiO_2.

You know how n-type semiconductors behave under these conditions.

The result of subordination:

> You can reasonably predict the photoconductive effect of ZnO and TiO_2 *(if/because)* you know how n-type semiconductors behave under these conditions.

As you can see, the subordinate strategy establishes a kind of ranking order between sentences. One sentence conveys the primary meaning of the statement, and the other sentences are used to qualify or modify the meaning of the base statement. In one sense, the added sentences depend on the main sentence for their meaning. You might say that the main sentence contains the first level of meaning, and the other subordinate sentences contain the second, third, and other levels.

Here is another example, with a *because* addition to the base:

The residents of the area were outraged.

The report did not accurately represent the dioxin levels in the soil.

Combined:

> The residents of the area were outraged *because* the report did not accurately represent the dioxin levels in the soil.

Try the next one yourself, using a *when* or *because* connector:

The engineers were shocked.

The report revealed that the building had been made with substandard material.

Preposition/Deletion Strategy

When we connect entire sentences there is a single operation—addition. But it's also possible to have deletion as a single operation. The result can be shorter sentences. When done well this operation can streamline flabby writing and result in conciseness and economy.

The key in this kind of deletion operation is the preposition. With prepositions, you can reduce clauses to simpler, more concise prepositional constructions. *Preposition/deletion* strategy, thus, goes one step further toward economy. Consider the following examples:

1. Since 1.5 liter polyethylene containers had been used, ...
2. Due to the use of 1.5 liter polyethylene containers, ...

Example 1 is a subordinate *since* clause, whereas example 2 is made up of two prepositional phrases. The prepositions are *due to* and *of*, and the rest of the words conclude the prepositional phrases.

To learn how to make preposition deletions as functionally as possible, it may help to review briefly what they are and how they are used. Prepositions act as connectors between their objects and the rest of the sentence. They indicate logical relationships such as the following:

time:	*at* 3:45 Greenwich mean time
place:	*in* the centrifuge
reason:	*for* electrolysis
manner:	*with* catalytic agents
possession:	*of* the biphenol group

Other commonly used prepositions are:

about	between	despite	over	through	without
against	because of	from	on	until	
before	by	like	to	with	

Looking over these columns will give you a good idea of what prepositions are and of the kinds of relationships they signal.

Now that you understand prepositions, let's examine the deletion operation in which prepositions are used to reduce subordinate clauses to phrases. Let's say that we've used a subordinate strategy to produce the following sentence:

> The residents of the area were outraged *because* the report did not accurately represent the dioxin levels in the soil.

To make the sentence even more economical, use a preposition/deletion strategy. Instead of the connector word *because*, use the preposition *because of* and delete the verb *represent* to come up with the following shorter sentence:

> The residents of the area were outraged *because of* the inaccuracy in the report on dioxin levels in the soil.

Combining with Miscellaneous Addition/Deletion Strategies · 51

In effect, what you have done is to begin by combining sentences with a subordinate strategy, and now you have further reduced the subordinate strategy to a more economical construction using a preposition. As you can see, the object of the preposition/deletion strategy is to reduce clauses to more economical constructions.

Here is another example:

The astronomers were annoyed *when* the students answered that Disney had discovered black holes in space.

The astronomers were annoyed by the students' answer that Disney had discovered black holes in space.

Try combining the following two sentences using a preposition/deletion strategy. See if you can work out a combination for yourself before going on to the suggested one:

Computers have become affordable enough for home use.

This is the result of inexpensive mass production of cheap silicon microchips.

Combined:

Computers have become affordable enough for home use

(because of)   inexpensive mass production of cheap silicon
(thanks to)   microchips.
(with the)

A number of prepositions might be used in this sentence; each one, of course, produces a slight difference in meaning.

Let's look at one more illustration. It first combines the sentences with a subordination strategy to produce a clause addition and then reduces the clause with a preposition/deletion strategy to achieve still more economy:

Interferon helps the body combat disease.

This man-made substance works like the body's own immunization system.

Combined to a clause:

> Interferon helps the body combat disease *because* this man-made substance works like the body's own immunization system.

Combined to a phrase:

> Interferon helps the body combat disease *by* working like the body's own immunization system.

It's not necessary to reduce all clauses to phrases. A rule of thumb is that if you can reduce the length of a sentence and simplify its structure, then go ahead and use a preposition/deletion operation. If the preposition produces a confused, meandering sentence, however, then don't use this strategy.

The procedure of reducing clauses to phrases can best be described with that word *tinkering*. A step-by-step outline follows:

1. Identify the clause, introduced by words such as *when, because,* and *since*.
2. Get a fix on the underlying meaning of the clause and consider what would be an appropriate preposition. (Usually, it will be one of the ones in the previous list.)
3. Eliminate the verb in the clause.
4. Add the preposition.
5. Make whatever other changes are necessary for sense.

TRIAL RUN EXERCISES

In the following exercises, combine each group of sentences into an economical, clear sentence, using, where possible, any one of the miscellaneous addition/deletion strategies that you have studied in this unit. Try to reduce the total number of words without losing any information. Where you use the subordinate strategy, try to economize further by using the preposition/deletion strategy.

1. Each company was contacted.
   Each division was recontacted.
   The recontacting was by telephone.

Combining with Miscellaneous Addition/Deletion Strategies · 53

    The recontacting was to answer any questions.
    The recontacting was to confirm participation.

2. It is the operator's duty to feed the stock.
   The feeding is on the bottom die.
   It is the operator's duty to see that the stock is properly positioned.
   It is the operator's duty to activate the press cycle.
   The press cycle is activated by pressing the foot switch.

3. The thermocouple protects homeowners.
   The protection is against explosion.
   The thermocouple is installed properly. (Try an *if* connector.)

4. We can keep the angular error within low values.
   The error is of the tone arm.
   The values are acceptable.
   The values are through the path of the stylus.
   The stylus is on the record.
   The tone arm is shaped.
   The shaping is suitable.

5. The amplifier and loudspeaker are combined.
   The combining is into one unit.
   The combining is frequent.
   The amplifier and loudspeaker are usually kept separate.
   The separation is for the purpose of reproduction.
   The reproduction is of sound.
   The sound is high fidelity.

6. Fluctuation in the speed manifests itself in "wow" and "flutter."
   The speed is of the turntable.
   The fluctuation exceeds approximately .003 of the speed.

7. The power press creates maximum hazards.
   The power press is used in high-speed manufacturing.
   The operator's hands move in between the dies.

8. The radiologist can use a contrast medium.
   The contrast medium is usually barium sulfate.
   The radiologist wants to examine internal organs.
   The examination is by X-ray.

9. The X-ray examination of internal organs is direct.
   The examination of internal organs is "live."
   The examination of internal organs is on a fluorescent screen.
   The technique is called fluoroscopy.
   The technique is used.

10. The house jitney driver has responsibility.
    The outside drayman has responsibility.
    The responsibility is for the cargo.
    The cargo is in transit between terminal areas.

COMBINING IN CONTEXT

Combine the following sentences into compact, economical prose. Where appropriate, use any of the combining strategies you have learned up to this point.

Adaptive Reuse of Historic Railroad Stations[1]

1. This study has resulted in a number of findings.
   The findings are in regard to projects.
   The projects involve reuse of railroad stations.
   It is adaptive use.
   The use will be in the future.

2. Railroad passenger service continues to decline. (Try an *as* connector.)
   It is declining on all rail corridors in the country.
   There are few exceptions.
   The need to develop creative new uses becomes increasingly urgent.
   The new uses are for the old passenger terminals.

3. The adaptive reuse of historic stations can be economical.
   It can be feasible.
   It can be a sound investment.
   The reuse is for a variety of new uses.

4. An objective is accomplished at the same time.
   The reuse accomplishes a major objective.
   The objective is to rehabilitate a structure.
   The objective is to preserve a structure.
   The structure is historic.
   The structure is culturally significant.
   The action preserves and rehabilitates the structure as a cultural resource.

5. The concept of adaptive use has been endorsed by local governments.
   It has been endorsed by citizens.
   It has been strongly endorsed.
   Governments and citizens support enterprises.
   The enterprises are commercial or cultural.
   The enterprises have been part of this endeavor.

6. The availability of stations is creating a growing need.
   Desire of public and private groups is creating a growing need.
   The stations are available for reuse.
   The desire of the groups is to undertake innovative reuse projects.

---

[1]Adapted from *Recycling Historic Railroad Stations* (Washington, D.C.: U.S. Department of Transportation, 1978), pp. 1–2, 53.

The need is for federal funding.
The need is for technical assistance to support these projects.

7. The need for federal funding and technical assistance is emphasized.
Rehabilitation and development are complex. (Try a *because* connector.)
And obtaining conventional financing is difficult.
The financing is for large-scale reuse projects.

8. The eight reuse cases illustrate specific strengths as well as pitfalls.
They illustrate these general findings as well.
These pitfalls were encountered.
These pitfalls serve to provide helpful advice.
The advice is for future reuse projects.

9. The difficulty of obtaining financing was among the obstacles.
These obstacles were noted most often.
The financing was reasonable.
The financing was for a project.

10. The various funding techniques serve as models.
The techniques were developed.
They are models for similar creativity.
They are models for ingenuity in future projects.

11. Another problem was an inability.
The problem was frequently noted.
It was an inability to generate income.
The income was adequate.
It was income to cover operating expenses.
This was once the project was completed.

12. A lesson was gained from stations.
The stations had a reuse plan.
The reuse plan was based on market analyses.
The analyses should be sound.
The plan should be based on economic analyses.
The analyses should include carefully compiled projected earnings.
The analyses should include expense statements.

13. Techniques provide useful examples.
They are sometimes used.
They are used to control annual expenses.
They are expenses such as leases.
The leases are where the tenant pays.
He pays for his own utilities.

Combining with Miscellaneous Addition/Deletion Strategies • 57

14. An obstacle has often been an inability.
    It is a third major obstacle.
    It has been the recycled station's inability.
    It is an inability to design facilities successfully.
    The designing is to accommodate needs.
    They are transportation needs.
    One need particularly is parking.

15. Several of the case studies show. (Try an *as* connector.)
    Transportation services are a use for the stations.
    They are often a logical use.
    They are often a compatible use.
    They are combined with civic or commercial functions. (Try a *when* connector.)

16. Planning can include facilities.
    The facilities are for parking.
    It is adequate parking.
    Planning can include traffic patterns.
    The patterns are for pedestrians.
    The patterns are for vehicular circulation.
    The patterns will lessen potential conflicts.

17. The New London Union Station project provides an example.
    It is a particularly illustrative example.
    Facilities for different modes of transportation are blended.

18. There are many obstacles. (Try the preposition *despite*.)
    Individual projects have been faced with these obstacles.
    The eight examples are proof.
    The proof is that nearly any obstacle can be overcome.
    It can be overcome with proper planning and persistence.

19. And the recycling of historical railroad stations is an experience.
    It is a rich and rewarding experience.
    The experience is for station developers.
    The experience is for the community.

Ann Arbor Circulation Plan for Future Traffic[2]

1. This plan outlines an approach.
   The approach is integrated.
   The approach is to the maintenance.

[2]Adapted from *Recycling Historic Railroad Stations*, p. 65.

It is maintenance of access.
The access is to the central area of Ann Arbor.

2. It deals with auto access.
It deals with parking.
It deals with pedestrian movement.
It deals with transit service.

3. Generally, the plan relies on use and service.
The use is more efficient.
It is use of transportation systems.
The systems are existing.
The plan relies on increased transit service.
The service is to accommodate demand.
The demand is growing.

4. It is recommended.
The bus service is to interface with commuter rail.
The bus service is high or intermediate level.

5. The agency would maintain the service.
The service is commuter rail service.
The service would include stops.
The stops are at the existing terminal.
The stops are at Dixboro Road.

Arc Welding[3]

1. Arc welding is a metal-joining process.
The process melts the base metal.
The process adds metal.
The metal results in a joint.
The joint is cooled. (Try a *when* connector.)
The process controls this through the use of electrical current and voltage.

2. This study is concerned with machines.
The machines are used to position, hold, weld, and eject arc-welded components.
The machines are used on a semi-production or full-production basis.

---

[3]Adapted from *Machine Guarding—Assessment of Need* (Washington, D.C.: U.S. Department of Health, Education, and Welfare, 1975), pp. 68–70.

Combining with Miscellaneous Addition/Deletion Strategies · 59

3. The operator may be a skilled worker.
   The operator may not be a skilled worker.

4. The process is semi-automatic. (Try an *if* connector.)
   The operator is usually a skilled worker.
   He has knowledge of welding processes.

5. It is the welder's responsibility.
   The welder must set up on the job.
   The welder must make machine adjustments.
   The welder must adjust and operate the welding head.
   And the welder must manage stock feeding.

6. The process is fully automated. (Try an *if* connector.)
   The operator may be a semiskilled worker.
   The operator may be stationed at a control panel.

7. It is in these cases.
   Machine setup and operation are performed by a millwright welder.
   They are performed initially.
   And they are then turned over to a production department.

8. This is in either case.
   The operator is remotely positioned.
   He is usually positioned away from the point of operation.

9. Hot sparks, extreme heat, and ultraviolet radiation are generated.
   This is at the point of operation.
   These can result in burns to the body.

10. The machinery will have clamping devices.
    The machinery will have stock-feed and eject mechanisms.
    The machinery will have head-feeding devices.
    These devices create pinch points.
    And these devices create in-running points.
    They may be hazardous.

11. The welding process uses electricity. (Try a *because* connector.)
    Improper controls can create shock hazards.
    Grounding can create shock hazards.

12. The point of operation must be closed.
    The welding operator must be provided with goggles.
    Or the welding operator must be provided with a shield.

13. Protective clothing is a must.
    The operator is near a welding zone. (Try an *if* connector.)

14. Mechanical devices must be enclosed.
    Or mechanical devices must be barricaded.
    This is to eliminate mechanical hazards.

15. Controls must be well maintained.
    Controls must be properly designed.
    This is to eliminate any electrical shock hazards.

16. Exhaust systems are desired to remove excess fumes.

Calibration and Standardization[4]

1. Take the methanol/ethanol mixture from a cylinder.
   This mixture is used as a standard.
   The cylinder is no more than 50 percent depleted.

2. This is done to assure the relative composition.
   The composition is of the components.
   The components are in the mixture.
   Studies have shown that the fraction may be high. (Try a *since* connector.)
   The highness may be abnormal.
   The fraction is of less volatile components.
   Eighty percent or more of the cylinder is depleted. (Try a *when* connector.)

3. Prepare a series of standards.
   Analyze a series of standards.
   The standards vary in concentration.
   The variance is over the range of 0–3000 ppm.
   The preparation and analysis are under constant temperature and pressure conditions.

---

[4]Adapted from *NIOSH Manual of Analytical Methods* (Washington, D.C.: U.S. Department of Health and Human Services, 1980), p. S–85–5.

Unit Six

Embedding Clauses and Phrases

Embedding, another basic operation, is an effective way of combining complicated units of meaning into a single sentence. This operation allows you to squeeze an entire sentence or phrase into one of the noun slots of a main clause. Embedding is a way to implant as much as an entire sentence into another by treating the implanted element as though it were a noun. The embedded sentence or phrase can go anywhere in a sentence where a noun might fit; it can be a subject or an object. We will discuss embedding on two levels. In the first, whole sentences (or clauses) are embedded, and in the second, *phrases* (that is, word groups without complete subject-verb elements) are embedded.

Embedding Clauses

Consider the following example, in which a whole sentence is embedded:

The experimenter concluded *something*.

Amplitude modification of microwaves has an effect on animal tissue.

Combined:

The experimenter concluded *that* amplitude modification of microwaves has an effect on animal tissue.

In this book, the words *something* or *this* in the main sentence will mark the spot where a sentence should be embedded. As you can see in the preceding example, the entire second sentence is embedded into the first. The word *that*, one of the connecting words in this basic operation, signals the start of an embedded element and also provides the grammatical glue that connects the sentences.

The word *that* is not the only connector used in this basic operation. For instance, *the fact that* is used to signal the embedding in the following example:

The noise levels exceeded the 90–dBA criterion curve set for this machinery.

Something had to be considered in the redesign of the equipment.

Combined:

The fact that the noise levels exceeded the 90–dBA criterion curve set for this machinery had to be considered in the redesign of the equipment.

In this example of embedding, the entire first sentence substitutes for a noun as the subject of the second sentence, and *the fact that* signals the start of the embedding and provides the grammatical link between the two sentences.

At times you will see a slight variation of the *that* connector in embedding operations. The word *it* is often used in combination with the word *that*, as in the following example:

It is clear *that* improved health insurance coverage is needed for workers in this industry.

Another connector used in embedding is the word *if*, sometimes followed by *or not:*

The ammeter will tell the technician *something*.

The ion concentration of the solution remains in the acceptable range.

Combined:

Embedding Clauses and Phrases · 63

The ammeter will tell the technician if the ion concentration of the solution remains in the acceptable range (or not).

Similar to the *if* connector is the *whether (or not)* connector:

The engineer had to determine *something*.

The octave-band spectrum measured at the filling machine inlet exceeded the 90-dBA level.

Combined:

The engineer had to determine *whether (or not)* the octave-band spectrum measured at the filling machine inlet exceeded the 90-dBA level.

Often questions—or sentences that could be put into question form—are embedded into nonquestion statements. The connectors in this case are question words such as *who(ever), whom(ever), which(ever), why, where(ever), when(ever),* and *how(ever)*.

Here is an example of embedding using *what:*

We need to know *something*.

What do toxic tolerance tests tell us about the residual effects in animal tissue of 10 parts per million of PCB in drinking water?

Combined:

We need to know *what* toxic tests tell us about the residual effects in animal tissue of 10 ppm of PCB in drinking water.

Question connectors can also be used to embed in sentences that are not in question form. For example:

The literature on this chlorobiphenyl isomer does not indicate *something*.

It must be used only in "closed systems."

Combined:

The literature on this chlorobiphenyl isomer does not indicate *why* it must be used only in closed systems.

The following examples illustrate embedding that uses other question connectors:

The Civil Aeronautics Board moved quickly to determine *something*.

How was the percentage of ethylene glycol solution changed between the time of its constitution and use?

Combined:

The CAB moved quickly to determine *how* the percentage of ethylene glycol solution was changed between the time of its constitution and use.

• • •

The CAB has set guidelines for temperature, humidity, and wind chill to indicate *something*.

When must the ethylene glycol mixture be raised above 70 percent ethylene glycol?

Combined:

The CAB has set temperature, humidity, and wind chill guidelines to indicate *when* the ethylene glycol mixture must be raised above 70 percent ethylene glycol.

• • •

The vibrating frequency of the quartz crystal and the input level of current determined *something*.

Which semiconductor switch design was going to be used in the servomechanism?

Combined:

The vibrating frequency of the quartz crystal and the input level of current determined *which* semiconductor switch design was going to be used in the servomechanism.

• • •

Biochemists want to know *something*.

Where in the blood are the enzymes found?

Combined:

Biochemists want to know *where* the enzymes are found in blood.

Embedding and Transforming Phrases

A remarkable thing about language is its flexibility. You can communicate the same message in a number of ways. Sometimes you merely change the form of a few of the words and thus change the grammatical function of those words. You can transform words and rearrange grammatical relations in a sentence, as in this illustration:

Words can be transformed, as you will see in this section.

This operation helps writers achieve economy of language.

Combined:

Transforming words, as you will see in this section, helps writers achieve economy of language.

In this example, we have changed the verb *transformed* in the first sentence into an *ing* noun, *transforming*. Then we have embedded the phrase containing the *ing* noun into the second sentence as its subject. *Transforming words*—this entire phrase—becomes the subject of *helps:*

Transforming words helps . . .

This procedure involves two basic steps, transforming and embedding. In addition, we have shortened the message; the combined sentence contains fewer words than the originals. So you can add the deletion operation to the list: transforming/embedding/deletion.

All of this maneuvering is meant to lead you to the observation that transforming involves multiple operations. But you have seen this phenomenon before, notably in the addition/deletion operation. Such complexity is to be expected, since the basic operations are, after all, the fundamental building blocks of sentence making. After you have practiced transforming in the following exercises, you will see that multiple operations come naturally, even instinctively, and do not depend on precise grammatical knowledge about the multiple operations. Practice and tinkering will allow you to master these operations.

But before moving to transforming and embedding, it may help to see in detail how transforming works. Verbs can be transformed into three different constructions:

1. A noun ending in *ing*
   oxidize ⟶ oxidiz*ing*
2. A *to* + verb form
   oxidize ⟶ *to* oxidize
3. A noun ending in *ion*
   oxidize ⟶ oxidat*ion*

The following examples show the three constructions plus embeddings.

1. Noun Ending in ing

Nuclei of light atoms such as hydrogen can be fused safely.

This has been accomplished in extremely strong magnetic fields.

Combined:

Fusing nuclei of light atoms such as hydrogen has been safely accomplished in extremely strong magnetic fields.

Verbs changed into nouns like this are called *nominalizations*. They abound in technical writing. While they can result in economy of language, they can also result in sentence structures that are very difficult to understand. They should be used with caution, but often are not. Since they are so frequently used in technical writing, and since they can reduce readability and comprehensibility of the text, Unit 9 will be devoted to their proper use.

2. To + Verb Form

Genes from different parent organisms have been spliced.

It is possible to do *this* because of discoveries in biochemistry and microbiology.

Combined:

It is possible *to splice* genes from different parent organisms because of discoveries in biochemistry and microbiology.

3. Noun Ending in ion

Nuclear wastes radiate damaging high-energy rays.

This has posed difficult engineering problems for modern technology.

Combined:

Nuclear waste *radiation* has posed difficult engineering problems for modern technology.

Transforming and Embedding with Prepositions

The nouns that are produced by transformations in examples 1, 2, and 3, like any other nouns, can be used as objects of prepositions. *Prepositions*, words that indicate logical relationships, act as connectors between their *objects* (the name given to nouns in prepositional phrases) and the rest of the sentence. Thus, by using prepositions with verbs transformed into nouns, the writer has available another strategy, this time one that takes advantage of the connecting power of prepositions as well as the economy and impact of transforming. Although this strategy is not used with great frequency, it is a handy tool. Here are some examples:

1. *Preposition* + ing *Noun:*

The noise level from the vibrating housing was reduced from 99-dBA to 92-dBA by *this*.

Styrofoam cushions were molded around the housing.

Combined:

The noise level from the vibrating housing was reduced *by molding* Styrofoam cushions around the housing.

2. *Preposition* + to + *Verb:*

For *this* energy must be supplied from the outside.

The energy level in any system remains constant.

Combined:

For any system *to remain* constant, energy must be supplied from outside the system.

3. *Preposition* + ion *Noun:*

Potential hazards of PCB in these industries result from *this*.

Fumes and dust containing PCB are inhaled.

Combined:

Potential hazards in these industries result *from inhalation* of fumes and dust containing PCB.

Incidentally, example 3 also could have been transformed into an *ing* noun with a preposition:

Potential hazards in these industries result *from inhaling* fumes and dust containing PCB.

Many *ion* and *ing* transformations can be made interchangeably.

A Review of Embedding and Transforming

As is true of sentence combining in general, you need not know all the grammatical terminology and concepts of embedding with great precision. If you keep a couple of things in mind, with some practice and tinkering, embedding will become almost instinctual. First, look for the signal *something;* in this book it indicates the point in a sentence where the embedding should be done. Remember the connectors *that, the fact that,* and *if,* as well as the question connectors *what, why, when, where,* and *who.* Finally, remember that an embedded sentence or phrase functions as the noun for which it substitutes, that is, as a subject or an object.

Unlike addition/deletion, embedding often does not result in fewer words than the original. The economy achieved in embedding comes from squeezing two or more sentences into one. This also allows you to pack more than one verb into the combined sentence. And since verbs, particularly when they denote action, provide the force and vitality of a sentence, embedding can, if done well, add punch to your writing.

Here is a review of the steps involved in transforming and embedding phrases. First, find the spot in one sentence where the noun substitution should take place, signaled here by the key words *this* or *something.* Second, locate the verb in the sentence that is to be transformed and reduced to a phrase. Transform it into either (1) an *ing* noun, (2) a *to* + verb noun, or (3) an *ion* noun. Make whatever other small changes are necessary, and put the transformed word and the rest of the phrase that accompanies it into the noun slot marked by the key word. If necessary for the meaning, use a preposition before the transformed word.

TRIAL RUN EXERCISES

Try embedding sentences in the following exercises. Let your ear help you decide which combinations sound best.

1. *Something* is surprising.
   The gyroscope mechanism will stay intact under pressures greater than five atmospheres.

2. No one knows *this*.
   When will the meeting be held?
   Where will the meeting be held?

3. The supervisor's immediate problem is *this*.
   How should he staff the project?

4. The state building inspector asked the contractor *something*.
   Will the substitute steel joints meet dead weight specifications?

5. There is no doubt about *this*.
   Reinforcing the structure will extend its life.

6. The research director was worried about *this*.
   Would the gene-splicing experiments produce uncontrollable bacteria?

7. The staff was told *this*.
   All personnel on the list would get an immediate raise and bonus.

8. He found *this*.
   What he learned in engineering class was valueless on the job.

9. *This* is remarkable.
   The TRS-80 has the flexibility to serve as both word processor and basic computer.

10. *Something* is clear.
    The inspector wants strict compliance with the code that states something.
    Contractors must use No. 12 copper romex.

• • •

In the following exercises, try some transformations and phrase embeddings on your own. Take the operations required one step at a time, and don't hesitate to tinker with the sentences until your ear tells you you have it right.

1. *This* is the only modification required.
   We must level the existing base.

2. Additional students can be entered into the data file by doing *this*.
   We must type a set of data cards.
   We must add them to the card deck.

3. *This* will cost time and money.
   The space will be filled with concrete.

4. We can eliminate the health problem by *this*.
   We drain the swamp.
   We install monitoring traps for mosquitoes.

5. *Something* is not feasible.
   We are to install chromatographic instruments.

6. We can insure proper pH level only by *this*.
   We can analyze water samples monthly.

7. *This* is acceptable to the board of directors of the airport.
   We will locate the runway on Halli Mountain.

8. The shop foreman's hardest task is *this*.
   He evaluates his employees' performance.

9. Accidents can be curtailed substantially by *this*.
   We will widen the shoulder in this area.

10. *This* will cost $1600.
    We will contract the consulting engineer.

COMBINING IN CONTEXT

Combine the following passages into complete paragraphs using the embedding operation whenever called for and any other operation that will produce smooth sentences. Use your own judgment to divide the sentences into paragraphs. Be economical with words, but don't leave out any information. Some groups of sentences contain key words; others do not.

Government Efforts to Control Health-Care Costs[1]

1. It was by 1986.
   Something had become clear.
   Improved health insurance coverage had also generated a wave.
   It was public and private insurance coverage.
   It was a wave of inflation in the health field.
   Meanwhile, improved insurance coverage had greatly increased the financial security of individual citizens.

2. It was as a result.
   A number of economists and conservatives called for a reduction.
   They were academic economists.
   They were political conservatives.
   The reduction was in the scope of coverage.
   It was public and private insurance coverage.

3. They argued *this*.
   Control of inflation required coverage.
   It was more coverage.
   It was not less coverage.
   It was health insurance coverage.

4. Their arguments were considered sound. (Try an *although* connector.)
   The soundness was technical.
   This approach was not politically viable.

5. Americans were not considered willing.
   The willingness was to return to payment.
   It was personal payment.
   It was the payment of medical bills.
   They were large and unpredictable.
   It was the price of slowing inflation. (Try an *as* connector.)
   Inflation is in the health sector.

6. The federal government's first choice was *this*.
   It called for restraints.
   The restraints were voluntary and self-imposed.
   The government was in search of a workable solution.

---

[1] Adapted from *Medical Technology: The Culprit Behind Health Care Costs* (Washington, D.C.: U.S. Department of Health, Education, and Welfare, 1977), pp. 16–17.

74 · *Style and Readability in Technical Writing*

7. These efforts included slight adjustments.
   They were adjustments in the design.
   It was the design of systems.
   They were private and public systems.
   They were health-insurance systems.

8. There was *this* in addition.
   Prominent individuals were asked to define and plan health needs.
   These individuals served in private agencies.
   The health needs were for their communities.
   They were designing and planning on a voluntary basis.

9. These voluntary efforts were largely ineffective.
   The inflationary pressures were *this* great. (Try a *so* connector as well as a *that* connector.)

10. Planning bodies were unable to make decisions.
    They were the kinds of decisions.
    The decisions were required for *this*.
    The decisions would curtail the spiral.
    The spiral was rapid and inflationary.

Technology and Health Costs[2]

1. The major studies have concentrated on the hospital sector only.
   The studies have shown *this*.
   Medical technology increases health costs.

2. This was noted above. (Try an *as* connector.)
   This could be a serious limitation.
   It could be a limitation to *this* extent.
   Analysis of hospital costs alone excludes benefits.
   Particularly these are per diem costs.
   They are technological benefits.
   These benefits avert the need.
   The need is for such care.

3. Nevertheless, hospitals account for 40 percent of health expenditures. (Try a *since* connector.)
   Hospitals are the factor.
   It is the health-care equation.

---

[2]Adapted from *Medical Technology: The Culprit Behind Health Care Costs*, pp. 32–33.

Hospitals are also the arena.
Most technologies are employed in the arena.

4. Studies have estimated the additional impact of technology.
The impact is on health-care spending.
The studies accounted for increased coverage.
The studies accounted for higher per capita income.
The studies accounted for greater availability.
The availability was of physician and hospital services.

5. Experts, however, would agree.
Much of the availability is attributable to *this*.
It is availability of medical care.
The medical care is "sophisticated."
It is costly.
Private and public insurance coverage is pervasive.

The Computer Tomography (CT) Scanner[3]

1. CT scanning is a procedure.
It is a diagnostic procedure.
It is a radiological procedure.
It is used for imaging mainly the head.
It is used for imaging the body also.

2. A CT scanner makes use of X-ray.
The X-ray is conventional.
The CT scanner collects information. (Try a *but* connector.)
It processes information.
It does this in a new way.

3. A source emits X-rays from positions.
There are several.
And a detector collects the energy.
It measures the energy.
The energy remains.
The X-rays have passed through a portion. (Try an *after* connector.)
It is a portion of the body.
The body is being scanned.

4. A computer constructs on a screen.
A computer displays on a screen.

[3]Adapted from *Medical Technology: The Culprit Behind Health Care Costs*, p. 116.

       It displays an image.
       It is an image of the area.
       The area is being scanned.
       All this is after *this*.
       All data are processed.

    5. A CT scanner makes a composite image. (Try a *because* connector.)
       The CT scanner has advantages over X-ray.
       The advantages are definite.
       The X-ray is conventional.

    6. *This* prevents overlapping organs from *this*.
       Narrow cross-sections are reconstructed.
       The organs obscure one another in images.

    7. CT scans also detect differences in density.
       They are small.
       The density is among adjacent structures.

    8. These two attributes make CT scans especially helpful for *this*.
       Low-density tissues can be seen.
       These are soft tissues.
       The brain is such a tissue.

    9. Tomographic devices were constructed in the United States.
       They were similar to the CT scanner.
       It was during the early 1960s.
       But they were not noticed by the medical community.

    10. The first CT scanner was developed in 1967.
        G. Hounsfield developed it.
        He is an engineer at Emitronics, Ltd. in Great Britain.

    11. It was 1971.
        The Department of Health supported *this*.
        It was the British department.
        The prototype scanner was installed.
        It was in London.
        And the Mayo Clinic installed the first scanner.
        It was in June 1973.
        In was in the United States.

    12. R. S. Ledley developed a model later.
        He was at Georgetown University Medical School.

It scanned not only the head.
It scanned the rest of the body.

13. It was marketed by Pfizer.
The first CT body scanner became operational.
It happened in early 1974.

14. It was thereafter.
This proceeded rapidly.
CT scanners were accepted.
CT scanners were diffused.

15. It was by May 1977.
At least 400 CT scanners were installed in the United States.

16. Machines are divided between head and body scanners. (Try an *although* connector.)
They are divided approximately equally.
Body scanners now account for most new purchases.

Forging and Hot-Metal Stamping[4]

Definition

1. *This* covers a large variety of machines.
Machines are hot-metal stamped and forged.

2. This study will focus on a process and on operations.
The process is forging.
The operations are similar to forging.

3. This is the forming of metal products by *this*.
Metals are forged and hot-metal stamped.

4. Forging and hot-metal stamping of metal products is *this*.
The temperature of the stock is elevated to increase its plasticity.

5. By *this*, forces are reduced and properties remain unchanged.
The metal is worked while hot.
The forces are forces of deformation.
The properties are mechanical.

---

[4]Adapted from *Machine Guarding—Assessment of Need* (Washington, D.C.: U.S. Department of Health, Education, and Welfare, 1975), pp. 83–85.

6. The process can be achieved through *this*.
   It is a forging process.
   Metals are rolled, hammered, upset, pressed, extruded, and drawn.

7. The basic process consists of *this*.
   The stock is heated to forging temperature (e.g., steel: 2000 degrees to 2300 degrees F.).
   The hot metal is placed into or between dies.
   It is struck and shaped.
   The flash is hot or cold trimmed.
   The metal is cooled.

Operator Involvement

1. It is in almost every forging operation.
   The operator must grasp the heated stock.
   He must grasp it with tongs.
   He must manipulate the stock on various die positions.
   The die positions are in the forging machine.
   He does this to shape it.
   He does this progressively.
   He does this shaping it into a finished product.

2. Controls are normally foot pedals.
   The controls are for operating the machinery.
   The foot pedals leave the operator's hands free.
   They are free to hold the stock.

3. The operator is closely involved in conditions.
   They are hot, noisy, and dangerous.
   They demand a high degree of safety orientation.

Hazards

1. The forging process is hazardous.
   It is one of the most hazardous operations in manufacturing.

2. The high level of hazard is inherent. (Try a *because of* connector.)
   Safety mindedness must be instilled. (Try a *because of* connector.)
   It is a high degree of safety mindedness.
   It is instilled in the operator.
   Thus, injuries are kept to a minimum.
   We compare this with similar cold-metal processes.

3. The common hazards include *these*.
   Extremely hot materials are handled.

Involvement is very close at the point of operation.
Other hazards include *these:* noise, accelerated work rate, machinery malfunction, slippery work stations, and airborne fumes and particles.

4. The operator must use tongs to handle the stock. (Try a *because* connector.)
   He keeps his hands clear of the zone.
   It is a metal-forming zone.

5. Power transmission equipment is normally enclosed.

6. Controls are simple foot pedals.
   The controls are for operating the machinery.
   They can be accidentally activated.

Guarding Need

1. Guards are not used.
   They are point-of-operation guards.
   They would hinder the operation. (Try a *because* connector.)

2. However, shields should be used.
   The shields are to protect against hot flying scale.

3. Power transmission components must be enclosed.

4. Operating controls must be designed.
   They are designed in the operation of machinery.

5. And a very high regard must be instilled.
   It is a regard for the inherent hazards.

An Overview of National Transportation Expenditures[5]

1. We even allow for an overstatement of transportation needs. (Try either an *if* connector or an *ing* addition)
   Something is evident.
   Any assessment will have a highway component.
   It will be an assessment of the future.
   It is the nation's future.

[5]Adapted from *National Transportation Overview* (Washington: D.C.: U.S. Department of Transportation, 1980), p. 75.

It is the nation's investment future.
The investment is in transportation.

2. Growth trends in an auto-oriented society are reversing. (Try *despite*.)
This makes it essential to do *this*.
The concrete infrastructure is deteriorating.
It is the infrastructure of our highways.
The deterioration is very rapid.
Expenditures must be planned.
The expenditures are considerable.
The expenditures are in the future.

3. *Something* is evident.
Investment would account for a substantial share.
The investment is in public transporation.
The public transportation is urban.
It is a share of investment.
It is investment in surface transportation.
This is especially true in the two largest groups.
They are population groups.
These groups surround large cities.

4. *Something* is also evident.
Per capita expenditure tends to increase.
It is expenditure for highways.
Plus, it is expenditure for urban public transportation.
The increase comes with the size of the urbanized area.
And larger per capita amounts would be spent.
They are significantly larger.
The spending would be in urbanized areas.
The areas have over a million population.

5. The proportional cost for rail transportation has not been included up to this point.

6. This might be expected. (Try an *as* connector.)
The percentage of rail cost varies.
It varies with population size.

7. *Something* was reported.
Rail was 71 percent of the total investment.
The investment was in urban public transportation.
This was overall.

Rail was reported to be 81 percent and 41 percent respectively. (Try a *while* connector.)
This is the case of the urban area with the largest population.
And this is the case of the urban area with the second largest population.

8. The expenditures present an interesting picture.
They are relative expenditures.
There is an expenditure for public transportation.
It includes the rail mode.
And there is an expenditure specifically for highways.
It's a picture of the two largest cities.
They are New York and Los Angeles.

9. New York would devote a high proportion of its funds.
They would be devoted to public transportation.
And New York would shift a large amount.
It would be shifted from highways to other public-transportation modes.
This would happen under a flexible grant system.

10. *This* is on the other hand.
Los Angeles would devote a relatively small proportion of its funds.
The funds would be devoted to public transportation.
Los Angeles would even make a slight reduction.
It would be a reduction in public transportation expenditures.

11. However, Los Angeles would not change *this*.
The city has allocated funds for highways.

12. Only three urban areas show large shifts.
The areas are New York, Boston, and San Francisco.
They are large shifts of funds.
The shifts are from highway to other public transportation expenditures.

13. These shifts are indeed substantial.

14. New York shifted nearly 2 billion.
New York is an example.

Unit Seven

Transposing Strategies

One of the things that makes English sentences flexible is that you can position elements in a number of places within a sentence: at the beginning, in the middle, between other units, and at the end. This flexibility makes possible stylistic variation as well as slight changes in emphasis and meaning. This unit deals with some of the guidelines for optional positioning of sentence elements. We call this operation *Transposing*. It occurs in combination with the addition operation.

For all practical purposes, we can limit our consideration of transposing to combinations involving (1) the noun/noun connection, (2) the ing/ed-connection, and (3) the subordination strategy. All these addition strategies produce meaning-units that can often be transposed in the sentence, as the examples that follow show.

For a noun/noun connection:

The cell, *a 16-cm-long and 13-mm I.D. windowless quartz tube,* is carefully wound with Nichrome heating wire.

Transposed:

A 16-cm-long and 13-mm I.D. windowless quartz tube, the cell is carefully wound with Nichrome heating wire.

For an ing/ed-connection:

Using charges of about 7,000 grams of resin and 1.5 liters of each solution, the laboratory prepares resin for product application.

Transposed:

The laboratory prepares resin for product-use application *using charges of about 7,000 grams of resin and 1.5 liters of each solution.*

For a subordination strategy:

When the 5 percent solution of HNO$_3$ is added to the effluent, a thick, white gel results.

Transposed:

A thick, white gel results *when the 5 percent solution of HNO$_3$ is added to the effluent.*

Since transforming involves moving elements from one place to another in a sentence, naming the positions in a sentence will help us discuss specific transposition strategies. We will designate additions that come before the main clause as left-branched and additions that come after the main clause as right-branched:

Left Branch-------------Main Clause-------------Right Branch

Here is an example of a left-branch transformation:

Using a thermocouple, the engineer designed a very precise thermometer.

Note that left-branch elements of more than a few words are followed by a comma.
Here is an example of a right-branch transformation:

The engineer designed a very precise thermometer using a thermocouple.

In addition to positioning modifiers in the right or left branch, you can also place modifiers between the subject and the verb, or in the *mid-*

dle of the sentence. Elements placed in the middle of a sentence are separated from the rest of the sentence by commas. In the following example, notice that such placement can make a sentence sound a little out of the ordinary:

> The engineer, using a thermocouple, designed a very precise thermometer.

Incidentally, this left, right, and middle terminology can be used to refer to any kind of modifier, including simple adjectives, adverbs, or prepositional phrases. In Unit 8, on readability, we will use this terminology in its widest meaning to show that transposing is a factor in readability.

Now that you have seen how transposing works, you need to see its uses. Basically, transposing can be used to achieve three objectives: (1) stylistic variety, (2) emphasis, and (3) continuity or transitions within and between sentences and paragraphs.

Although the importance of stylistic variety has probably been overstressed in textbooks, transposing can break the monotony when the same sentence patterns, used too often, begin to irritate the reader. For instance, short, choppy sentences that invariably begin with a subject and verb followed by a subordinate addition can be transposed so that a right-branch subordinate clause moves to the left-branch position. More important, a sentence style that is heavily left-branched, and therefore, as you will see in Unit 8, relatively harder to read, can be transposed to a predominately right-branched style.

The second use of transposing is to create or shift emphasis. Listen to the differences in emphasis in the following examples. (In addition, note that commas are used after the left-branch *if* and *to* clauses.)

1. A. If an isolator is too weak vertically, it may not be laterally stable.
   B. An isolator may not be laterally stable if it is too weak vertically.
2. A. To avoid this difficulty, we can use side restrained metal-spring isolators.
   B. We can use side restrained metal-spring isolators to avoid this difficulty.
3. A. If steel springs rest on elastomer pads, we see much improvement.

B. We see much improvement if steel springs rest on elastomer pads.

In example 1, version A emphasizes the vertical weakness of the isolator more than version B. In version B, lateral instability seems more emphatic than vertical weakness.

In example 2, version A stresses the idea of avoiding the vibration problem, whereas version B emphasizes the availability of metal-spring isolators to do the job.

In example 3, version A places more stress on the use of elastomer pads. Version B downplays the elastomer pads and gives predominance to the idea of improvement.

In examples 1, 2, and 3, we've changed emphasis, ever so subtly, just by transposing elements, moving subordinate additions from the left-branch to the right-branch position, and vice versa.

You probably think that right-branch elements get the stronger emphasis, and while this conclusion is not a bad generalization, it is not true in all cases. It would be more accurate to say that any addition to the main clause placed out of its more normal order draws attention to itself. Left-branch additions such as the subordinate strategy normally come after the word in the main clause that they modify. Placing one before the main clause, therefore, causes it to be more or less out of place and thus more emphatic.

But you don't really have to worry about how emphasis is changed; your ear will tell you which transposed version produces the emphasis you want, and that is what matters. Experiment with transposing until you get the results that convey your message most effectively. When you look at two or more versions of a sentence that has been transposed in different ways, you'll see that you probably prefer one over the others, a sure sign that your ear is selecting the version with the most appropriate emphasis.

Perhaps the most important benefit of transposing is to provide the reader with links and smooth transitions between chunks of material within sentences, as well as between sentences. Transposing offers you a way to create transitions in sentences by moving elements around so that key words and ideas are placed as close to each other as possible. Since readers read words in chunks, as well as sequentially, the close placement of key words establishes tie-lines between chunks of material.

The following diagram shows you how various addition strategies have been transposed to place key words or ideas close to each other:

The agency required a reduction of the elevator noise by at least 30 dB in this situation. Because the operation is automated, the design engineers considered [enclosing] the two units involved. Such a [treatment] would normally be considered routine. In this case, however, because a food processing facility is involved, the agency set rigid requirements to avoid contamination of the food products from acoustic infill particles used in the construction of the [enclosure] panels. In addition the [enclosure] had to accommodate product heat loss.

Let's consider the transition between the first and second sentences. The word *operation* in the second sentence refers back to the word *situation* in the first sentence. The operation is part of the equipment *situation*, which requires noise reduction. In the diagram, the boxes connected by lines contain the key words *situation* and *operation*. Appearing in a left-branch addition, the key word *operation* comes as close as possible to the other key word, *situation*, which is located at the end of the previous sentence.

In this schematic passage, we have transposed key words or ideas in strategic additions to a left-branch or right-branch position so that they are as close as possible to other key words. In short, optional placement through transposing, as indicated by the connected boxes in the diagram, makes it possible to put key ideas in proximity with one another in order to create transitions for the reader.

Keep in mind that while transposing gives you options, there are limits. Then, too, words, phrases, and clauses can be transposed to the wrong place, thus resulting in ambiguity. Misplaced elements can contribute to lack of clarity and distortion in meaning, as occurs in this example:

> Containing misplaced modifiers, the writer edited all the sentences in the paragraph.

Transposing the *ing* connection to the left-branch position just does not work in this sentence, which should read:

> The writer edited all the sentences in the paragraph containing misplaced modifiers.

Here is another silly sentence that can result from misplaced modification:

> Charged with 10,000 volts of electricity, the technician carefully worked among the electrodes.

The mistakes in these examples are easy to spot and will cause only a chuckle or two. But in the next example, potentially serious consequences could result from a misplaced modifier:

> Containing no more than 100 ppm of impurities, the benzoic acid is added to hydrogen sulfide.

Consider, for example, how the meaning of the following sentence is significantly different from the preceding one:

> The benzoic acid is added to hydrogen sulfide containing no more than 100 ppm impurities.

When there is any possibility of ambiguity, modifiers should be placed as close as possible to the element modified. If you remember this general rule, transposing won't give you any trouble.

TRIAL RUN EXERCISES

In the following exercises, first combine each group into sentences using any of the strategies already studied; then transpose each of the sentences you have written into at least two versions. Compare the relative effectiveness of each version in class so that you can listen to your colleagues and work out a consensus with regard to which versions are most effective.

1. The REX distilling method proved superior to the Candelari procedure.
   The REX distilling method used raw potatoes.

2. The barometer mercury fell .5 in./hr.
   The meteorologist noticed the sudden change in the barometer.
   He began plotting.

3. The process continues.
   It continues without operator involvement.
   The machine is actuated. (Try a *once* connector.)

4. Exposure is suspected. (Try an *if* connector.)
   It is exposure to xylene.
   It is exposure to salicylic acid.
   It is exposure to styrene.
   A method should be used.
   The method employs chromatographic separation.

5. The individual has ingested aspirin. (Try an *if* connector.)
   Or the individual has been exposed to styrene.
   The method should not be used.
   Metabolites of these compounds will react. (Try a *because* connector.)
   They will produce a colored product.

6. The loading checker removes the dock receipt.
   The receipt is from the cargo.
   The loading checker counts the freight.
   The counting is during the loading process.
   A break bulk shipment is loaded into a container. (Try a *when* connector.)

7. The low-pressure cell appeared on the hurricane map.
   The meteorologist began plotting the low-pressure cell.

8. Many studies have been made on *this*.
   How is the flow of water through pipes, orifices, and nozzles measured?
   It was during the two thousand years between 300 B.C. and A.D. 1700.
   But significant progress was not made.
   It was not made until the end.
   It was the end of that period.

9. Pitot tubes have several characteristics.
   The characteristics are advantageous for laboratory work.
   They are advantageous for certain special applications.
   Current meters have been found more practical for use. (Try an *although* connector.)
   They are meters of the conventional type.
   The use is on rivers.

10. The tube bent 90 degress on the end doubled the reading.
    It was the reading made on the scale.
    The tube bent 90 degrees on the end reduced probable errors.
    They were errors that entered into such readings.

COMBINING IN CONTEXT

Combine the following sentences using all the strategies you have learned previously: addition/deletion, wh-connection, noun/noun connection, ing/ed-connection, embedding, and transposing. Experiment with transposing the additions into right-branch, left-branch, and middle positions.

Maritime Cargo Control[1]

The Operation

1. A trucker arrives at the empty storage area.
   The trucker has proper authorization.
   It is authorization to pick up an empty container.

2. Yard personnel examine his truck pass.
   They do it to verify his authority.
   They do it to issue an empty container.

3. The yard personnel initial the truck pass.
   The drayman proceeds toward the gate.

4. The gate guard is responsible.
   He is responsible for security procedures.
   The security procedures are for the issuance of empty containers.

5. The truck pass serves as his operating document.

Empty Container Pickup

1. Terminal personnel verify the nature of the drayman's visit.
   Personnel verify his authority.
   It is authority to receive a particular container.
   This is done in order to release a container.

2. The terminal personnel initial the truck pass.
   This is done to indicate their approval of the pickup.
   The initialing comes after the container has been provided to the drayman.

3. Empty containers should be opened to do *this*.
   Personnel should make sure no loose cargo remains.

4. Full containers should be examined.
   The examination is for proper container number.

5. The procedures should be part of the gate guard's responsibilities.
   This is the case if the yard personnel are unable to perform the container-examination procedures.

---

[1] Adapted from *Maritime Cargo Loss Prevention*, Vol. 1 (Washington, D.C.: U.S. Department of Transportation, 1979), pp. 35–36.

Empty Container Moving Toward the Gate

1. There is virtually no possibility for theft.
   This is the case while an empty container is moving.

2. However, it is possible to place cargo in the box.
   It is stolen cargo.
   This is the case if a driver stops in the yard.
   The driver has an empty container.

3. The chance of *this* is minimized.
   The theft occurs.
   The minimization is when draymen are provided clear directions.
   These are directions upon *this*.
   The draymen leave the outbound area.

4. Directions can be provided orally and visually.
   The oral directions are by the yard personnel.
   The visual directions are through *this*.
   A container pass is used.
   The visual directions are through a well-marked container yard.

5. Yard personnel should be instructed to remain alert.
   The alertness is for draymen with empty containers.
   The containers are at rest in the yard.
   And they should periodically examine the container.
   It is a container of any drayman at standstill.

6. Crane operators have an ideal vantage point.
   Straddle carrier drivers have an ideal vantage point.
   From this point they can notice parked trucks.
   The crane operators are seated high above the ground.
   The straddle carrier drivers are seated high above the ground.

7. The drayman will stop.
   And the drayman will have his equipment tested.
   It will be tested for roadability.
   This testing will be at the inbound inspection station.
   The drayman will stop before he reaches the gate.

8. Yard personnel will examine container and chassis.
   They will be examined for any defects.

9. The container and chassis will be examined.
   And exceptions will be noted.

The examination is for any defects.
The exceptions will be noted on the equipment interchange.

10. Procedures for inspection usually have the mechanic do *this*.
    He enters the box.
    He looks for damage.
    The damage is on the top and sides.
    The inspection is of the empty container.

11. This provides an opportune time for *this*.
    The mechanic can perform a security function.
    He can also look for loose cargo. (Try a *because* connector.)

Handling of Samples[2]

Collection and Shipping of Samples

1. Attach a small piece of Teflon tubing.
   Attach it to the hose bib.
   It is the hose bib of the five-layer bag.
   It is a gas-sampling bag.
   This should be done immediately before sampling.

2. Do not use the rubber tubing.
   Air will pass through the pump and tubing.
   Air is sampled.
   Air passes through before entering the sampling bag.
   A push-type pump is required. (Try a *since* connector.)
   No tubing is attached.
   The attachment is to the inlet of the pump.

3. Set the flow rate.
   Do this as accurately as possible.
   You are to use the manufacturer's directions.
   It is necessary to do *something*.
   The volume of the sample must be kept to 80 percent or less.
   The percentage is of the bag's capacity.
   The volume of sample collected is not used. (Try an *although* connector.)
   The sample is collected.
   The use is for *this*.
   The concentration is determined.

[2]Adapted from *NIOSH Manual of Analytical Methods* (Washington, D.C.: U.S. Department of Health and Human Services, 1980), p. S-85-5.

4. Record the temperature and pressure.
   It is the temperature of the atmosphere.
   It is the pressure of the atmosphere.
   The atmosphere is being sampled.
   Record the elevation.
   The pressure reading is not available. (Try a subordinate connector.)
   Also record sampling time.
   Record flow rate.
   Record type.
   It is the type of sampling pump.
   The pump is being used.

Analysis of Samples
1. Attach the gas sampling bag to the loop.
   It is the sample loop of the chromatograph.
   Do this with a piece of tubing.
   The piece is short.
   The tubing is flexible.

2. Open the valve.
   It is the valve of the bag.

3. Fill the loop by *this*.
   A vacuum pump should be used.
   Or manual pressure should be applied to the bag.

4. Allow the loop to attain pressure.
   It is atmospheric pressure.

5. Inject the sample.
   Make duplicate injections of each sample.
   Make duplicate injections of each standard.

6. No more than a 3 percent difference is to be expected.
   The difference is in area.

Two Kinds of Handsaws
1. You probably know *something*.
   The handsaw consists of two parts.
   The handsaw is the common handsaw.
   The two parts are a blade and a handle.

2. However, you might not know *something*.
   Handsaws come in two varieties.

The varieties are blade varieties.
One variety is the crosscut.
And one variety is the ripsaw.

3. The crosscut blade is used.
   The using is to cut.
   The cutting is across the grain.
   The grain is of a piece of wood stock.

4. The ripsaw is used.
   The using is "on the other hand."
   The using is to cut.
   The cutting is with the grain.

5. In other words, you would use a crosscut blade.
   The using would be to saw.
   The sawing would be across the width.
   The width is of a board.

6. You would use a ripsaw.
   The using would be to saw.
   The sawing would be of a lengthwise strip.
   The lengthwise strip would be from a board.

7. As a home handyman, you should learn.
   The learning is of the difference.
   The difference is between the blades.
   The blades are the two blades.

8. You should also know *something*.
   Each blade works as it does. (Try a *why* connector.)

9. A quick look will show you *something*.
   The blades differ. (Try a *how* connector.)
   The looking is at the teeth.
   The teeth are of the two saws.

10. The teeth are bent.
    The teeth are on both types.
    The bending is outward.
    The bending is from the blade.

11. One tooth bends to the left, the other to the right.
    The bending is alternately.

The bending is down the length.
The length is of the blade.

12. Hold the saw.
    The holding is to examine the teeth.
    The holding is horizontally.
    The holding is with the cutting edge up.

13. Each tooth is bent in an opposite direction.
    Something will give the illusion.
    The illusion is of two rows.
    The rows are of teeth. (Try a *the fact that* connector.)

14. Don't be distracted.
    The distracting is by this.

15. Concentrate on *something*.
    Each tooth is shaped.
    And each tooth is sharpened.

16. The teeth will be sharpened.
    And the teeth will come to tips at the outer edges.
    The teeth are on a crosscut blade.
    The sharpening is at an angle of about 65 degrees.
    The tips are knifelike tips.

17. Ripsaw teeth are sharpened.
    The sharpening is at 90 degrees.
    Or the sharpening is straight across.

18. They are square.
    They look much like chisels.
    The chisels are tiny chisels.

19. The knifelike teeth provide action.
    The teeth are of the crosscut.
    The action is a cutting action.
    Or the action is a slicing action.

20. This action is *something*.
    It is needed for cutting.
    The cutting is across the grain.
    The grain is of wood stock. (Try a *what* connector.)

21. The teeth of the ripsaw work.
    The teeth are square teeth.
    The working is like a series.
    The series are of fine chisels.

22. In a loose sense, the ripsaw chisels.
    Or, in a loose sense, the ripsaw rips.
    The chiseling or ripping is instead of cutting.

23. This chiseling works effectively.
    Or this ripping works effectively.
    The working effectively is when sawing.
    The sawing is with the grain.

Transportation and the Elderly[3]

1. The year was 1980.
   Twenty-five million Americans were over the age.
   The age was 65.
   About 62 percent of these lived in areas.
   The areas were urban or suburban.

2. We estimate *this*.
   There will be 32 million over 65.
   This will be by the year 2000.

3. Only about one-fifth have employment.
   The one-fifth is of the elderly.
   The employment is of any kind.

4. Their average income is roughly half.
   It is half of that of families.
   The families are of those under 65 years of age.

5. Almost 5 million live in households.
   The 5 million are of the elderly.
   The households have incomes below the level. (Try a *with* connector.)
   The level is a poverty level.

---

[3]Adapted from *1972 National Transportation Overview* (Washington, D.C.: U.S. Department of Transportation, July 1972), p. 173.

6. Transportation is their expenditure.
   The expenditure is their third highest one.
   It takes ten cents out of every dollar.
   The dollars are the ones they receive.

7. The elderly drive much less.
   The driving is less than other segments.
   The segments are of the population.

8. Transportation is expensive. (Try a *because* connector.)
   The transportation is privately owned.
   Only a small percentage can afford automobiles.
   The percentage is of the elderly.
   The elderly often have limited incomes.

9. The elderly experience some problems.
   The problems are common ones.
   The common ones are in the use of transportation.
   The transportation is public.

10. Design features act to their disadvantage.
    The design features are such as these.
    There are high entrance steps.
    There are overhead grips.
    There are fast-acting doors.

11. Studies have shown *this*.
    Waiting stations cause discomfort.
    Waiting stations cause a health danger.
    The waiting stations are unsheltered.
    The discomfort and health danger are under conditions.
    The conditions are severe weather conditions.

12. *This* is more confusing to the elderly than to younger groups.
    Schedules and routes are complex.

13. *This* adds to their fatigue.
    This increases the danger.
    The danger is of exposure.
    The exposure is to criminal assaults.
    The exposure is to inclement weather.

14. All these factors tend to inhibit *this*.
    The elderly use public transportation.

The elderly are the major users of such systems.
The systems are for *this*.
The elderly shop.
The elderly visit friends and relatives.
The elderly attend church.
The elderly keep doctors' and dentists' appointments.
The elderly engage in social and recreational activities.

Transportation and Air Pollution[4]

1. No one questions *this*.
   The nation's transportation systems are a source.
   The source is a major one.
   The source is of certain kinds of air pollution.

2. The EPA estimates *this*.
   Seventy-three and six-tenths percent of emissions came from sources.
   They were transportation sources.
   The emissions were carbon monoxide.
   It was in 1979.

3. Transportation also contributed to 52.9 percent.
   The percentage was of hydrocarbons.
   Transportation contributed to 47.1 percent.
   The percentage was of nitrous oxide.

4. Research data show *this*.
   The vehicle was the primary source.
   The vehicle was powered by the gasoline engine.

5. The source was of carbon monoxide (86.6 percent).
   The source was of hydrocarbons (85.4 percent).
   The source was of nitrous oxide (67.9 percent).

6. Nonhighway use of motor fuel was the other source.
   It was a major source.
   The source was of these three pollutants.

7. The various pollutants are in no way equal.
   The equality is in their effects.
   The effects are on health.

[4]Adapted from *1972 National Transportation Overview*, p. 171.

The effects are on property values.
The effects are on vegetation, and so on.

8. *This* happens at a given concentration.
   Some pollutants are more toxic than others.
   Some pollutants are more unpleasant than others.
   The toxicity and unpleasantness are at given concentrations.

9. It is for this reason.
   The EPA's 1981 standards establish different levels.
   The standards are ambient air quality ones.
   The levels are tolerable levels.
   The levels are for each type of pollutant.
   The levels are in accordance with the toxicity.
   The toxicity is relative.
   The relativity is to each type of pollutant.

10. Transportation still remains a major source.
    The source is of pollution.
    But the data indicate *this*.
    Transportation emissions are quite high. (Try a *while* connector.)
    The emissions are carbon monoxide.
    Carbon monoxide is believed to be one of the pollutants.
    It is one of the less toxic pollutants.

11. It is nationwide.
    Transportation causes about 16.7 percent of all pollution.
    The percentage is somewhat less than the percentage. (Try a noun/noun connection.)
    The percentage is attributed to power sources.
    The percentage is attributed to industrial sources.

12. Yet it is by the same measure.
    Transportation produces almost 70 percent of the pollution.
    The pollution is in Los Angeles and San Diego.
    Transportation produces over 50 percent on the average.
    The average is in selected areas.
    The areas are urban.
    The areas are selected throughout the United States.

13. Thus transportation pollution is important.
    Where does it occur? (Try a *because of* connector.)
    What amount does it occur in? (Try a *because of* connector.)

Unit Eight

Readability Guidelines

We've all struggled with material that's hard to read. At times the subject matter itself causes the problem, but sometimes it's the writer who is the problem. A poor writer can make simple material difficult, and a skilled writer can make difficult material easy.

But how does a writer make any material as easy to read as possible? That's the question we'll address in Units 8, 9, and 10. First, in this unit, we'll explain guidelines for producing sentences that are easily readable. We'll say a word about how to analyze an audience to determine the level of reading difficulty that's right for it. Then, in Unit 9, we'll look at a widely used type of noun that is a particularly important factor in readability. Finally, in Unit 10, we'll turn our attention to techniques for achieving readability.

No audience, however sophisticated, likes to have to stretch to the limit of its reading capacity. And many audiences are fairly limited in their capacity to handle difficult writing, anyway. This means that as writers, especially as writers of functional, technical material, you should always keep an eye on the factors that make sentences not only accurate and correct but also as readable as possible. Although what makes sentences and passages readable is a complicated matter, research has isolated a dozen or so of the most important things to consider.[1]

Here are ten guidelines for improving the readability of sentences.

---

[1] We are heavily indebted to the Document Design Center of Washington, D.C., for their excellent research and publications on readability.

1. As Often As Possible, Make the Subject of the Sentence the Doer of the Action that Is Contained in the Verb

Consider this sentence:

Field technicians tested core samples on site.

The technicians are doing the action in the verb; they're testing. Now consider this example, in which the subject is acted upon:

Core samples were tested on site by field technicians.

In the second sentence, *core samples* aren't doing anything. They are neither the doers nor actors of the action contained in the verb. Obviously, *core samples* are not testing.

Is Verbs

What about verbs such as *is, are, was, were, has been,* and *had been?* Most language theorists think there is no action in these verbs. Thus if we follow rule 1, we would avoid them altogether. But these verbs play such an important role that we can't do without them. So don't try. If the concentration of them gets too high, however, the message will become fuzzy. So, whenever you can, substitute an action-packed verb for an *is* verb, especially in a crucial spot. Look at this sentence, containing an *is* verb:

But these verbs *are* so important that we can't do without them.

Revised to eliminate the *is* verb the sentence reads:

But these verbs play such an important role that we can't do without them.

Dummy Subjects

The words *it* and *there,* used with *is* verbs, often act as "dummy subjects," creating sentences that cannot have actor or doer subjects. For example:

It is clear that 50 lbs./in.² is too much pressure for the seals to withstand.

There are eight factors to consider when calculating the stress on the wing struts.

These sentences are not hard to read, but if they are part of a long string of sentences with actorless subjects, the readability of the passage will decrease. These particular sentences in the examples can be easily revised, as follows:

Clearly, 50 lbs./in.² exceeds the pressure limit of the seals.

You (or the engineer) must consider eight factors when calculating the stress on the wing struts.

Don't try to eliminate "dummy subjects" altogether. They are helpful at times. But do use them sparingly.

Passive Voice

Overuse of the passive voice also can cause a problem. Passive-voice sentences contain subjects that are in some way acted upon. The action of the verb is done *to* the subjects, as this example shows:

1. All data were subjected to a standard co-variance analysis for verification.

The verb *were subjected* is in the passive voice. The subject is being acted upon and, in this sense, is passive. This passive-voice sentence can easily be changed into the active-voice sentence with a doer subject:

2. For verification, experimenters subjected all data to a standard co-variance analysis.

Sentences 1 and 2 are both relatively clear, but according to readability experts, sentence 1 (passive) takes longer to comprehend than sentence 2 (active), even for a good reader. When you pile passive voice, actorless sentences on top of one another, you compound the reading difficulty that would not be troublesome if only one or two sentences were involved. To prove the point, here is example 1 along with other passive-voice sentences. Perhaps you can hear the reading difficulty begin to build up:

Samples were collected systematically for over a year, and data were recorded according to the Fritz method. All data were subjected to a standard co-variance analysis for verification. Calibration of instruments was conducted prior to sample gathering and repeated to determine maintenance of accuracy.

Now look at these same sentences reconstructed into the active voice:

We collected samples systematically for over a year and recorded data according to the Fritz method. We subjected all data to a standard co-variance analysis for verification. We calibrated instruments prior to sample gathering and repeated the calibrations to double-check their accuracy.

The passive voice is the main problem in this passage. But there is another problem that we will only mention here because Unit 9 will be devoted to it later. It is the problem caused by abstract words such as *calibration*. The more abstract a word is, the harder it is for the reader to imagine it as the doer of any action. Such abstract words usually end in *ion, ance,* or *ence*. Even when these nouns are not in the subject position, too many of them can give the reader real difficulty.

When you run across a difficult sentence with either a passive-voice verb or an abstract noun for a subject, here's the way to revise it:

1. Locate the most significant action in the sentence, perhaps, but not necessarily, found in the verb. The action may be implicit.
2. Ask who is doing the action.
3. Then make the actor or doer of that action the subject of the verb, whether the doer is in the original sentence or not.

Let's revise one more passive-voice sentence using the three steps:

Maintenance of hydraulic equipment must be monitored regularly by shop supervisors.

What is most significant action? Monitoring. Who is doing the action? Supervisors. The revision would read like this:

Shop supervisors must regularly monitor the maintenance of hydraulic equipment.

A word of caution. Don't try to drive the passive voice into extinction. It is useful when you want to focus on the object of the action of the verb in the sentence or on a process that is being done. In fact, research shows that in some contexts, passive-voice sentences are as easy or easier to read than their active-voice counterparts, particularly when you want to spotlight the action in the verb rather than in the doer of the action. Here's an example:

> The tubes are constructed of glass tubing with both ends sealed.

You certainly aren't interested in who made the tubes; you're interested in what they're constructed of.

2. Use Personal Pronouns

Personal pronouns such as *I, you, he, she,* and *they* can help you remove the stiffness and impersonality from writing. Researchers have found that personal pronouns can improve readability. Often, they are a good cure for the overuse of the passive voice. Here is an example:

> The diskette containing the word-processing program must be removed from the disk drive before the editing commands can be given and the text entered into the Apricot III microprocessor.

Let's revise this, using the pronoun *you*, a particularly appropriate word since it is a commonly used pronoun in written directions and other technical writing:

> You must remove the diskette containing the word-processing program from the disk drive before you give editing commands and before you enter text into the Apricot III microprocessor.

A word of caution: Make sure the reference for a pronoun is clear and consistent throughout the text.

3. Avoid Difficult Words When You Can

It's hard to say when a word is difficult, because what is difficult to one person may be familiar to another. But we all have a pretty reliable inner sense that tells us when the words we use might be too tough for the audience. We should rely on this sense. But there are other things to keep in mind. For instance, the more abstract a word is, the more difficult it is for readers to process through the language centers of their brains. When you must use abstractions, also include analogies to concrete things. They will give the audience a familiar frame of reference. Here's an illustration:

> The calcium ion remains suspended in the bloodstream and acts like a trap to catch and block the microelectrical impulses that are aimed at the nerve centers in the ventricle.

Discussions of ions and electricity are abstract, hard to visualize, but the words *trap, block,* and *aim* are simple, concrete action words that give the reader a visual frame of reference. Many nontechnical magazines and newspapers do a good job of giving the reader concrete analogies (even in their science and technology sections).

Latin vs. Anglo-Saxon

Because the vocabulary of modern English has a Latin as well as an Anglo-Saxon origin, English speakers often have several words to say the same thing. Generally, a Latinate version is more abstract sounding, longer, and harder to understand than the Anglo-Saxon version. Consider these illustrations. Which is simpler?

> The committee *effectuated* the plans of the production team.

> The committee *carried out* the plans of the production team.

There's no need to tell you which sentence contains the Latinate version. The following lists should give you an even better basis for comparison:

Latinate	*Anglo-Saxon*
abrogate	give up
aggregate	group or collect
promulgate	publish or start
terminate	end
utilize	use

Redundancy

Redundancy often comes from the overuse of Latinate words. Writers who like Latin-based words are often trying to sound very official. They are usually the same people who pad sentences with phrases like "first and foremost," "general consensus of opinion," "oftentimes," "extraordinarily unique," and so on. Avoid these. Use economy. Remember the deletion operation. Combine sentences and cut out flab.

4. Avoid Stringing a Cluster of Nouns Together Without Connector Words in Between to Glue Them Together

Technical writers have become accustomed to noun clusters. But researchers report that they're hard to read, particularly when there are more than two in a row. Here are some examples:

1. blood cholesterol maintenance control diet
2. chemical environment protection agency

When you cluster nouns in this way, you're trying to make them function as adjectives, but without using the necessary connective prepositions such as *in, of, for,* or *about.* If we put those connectors in, example 1 might look something like this:

1. a diet to control and maintain cholesterol in blood

This example makes sense, but look at example 2 after translation:

2. agency for monitoring the protection of the chemical environment

Or should it read:

2. agency for protecting the environment and monitoring chemical abuse

No one knows. That's the trouble with noun clusters. Avoid them.

5. Avoid Deletion of Wh-Connectors in Some Wh-Connections

Sometimes we create an ambiguous sentence when we leave out the *who* or *which* in wh-connections:

> The supervisor wanted the chemicals locked up in low-humidity storage.

This could mean one of two things:

1. The supervisor wanted the chemicals that were locked up in low-humidity storage.

Or:

2. The supervisor wanted someone to lock up the chemicals in low-humidity storage.

If you mean example 1, inserting the wh-connector *that* will make your meaning clear. Some wh-connectors, however, can be deleted without causing ambiguity. Always double-check to make sure.

6. Do Not Transpose Lengthy Additions Between Subject and Verb or Between Verb and Object

Avoid putting long interruptions between the main elements of a main clause. Research shows that such interruptions impede the flow of meaning and result in hard-to-read sentences. Look at this example:

> The task force, on engineering grounds, on conservation grounds, and on economic grounds, approved all design details.

Too much information separates the subject from the verb. This sentence can be revised to take care of the excessive additions to the middle position:

> The task force approved all the design details because they met engineering, conservation, and economic standards.

Transposing and revising are the best ways to improving the readability of sentences with excessive medial additions.

7. Remove Multiple Conditions from Left-Branch or Middle Positions

Technical writing contains a lot of *if* clauses, which state conditions that have to be followed in order to achieve certain results. The more of these you put in a sentence, the harder the sentence will be to read. You can solve this problem by revising the sentence so that you can list the conditions separately—preferably in the right-branch position of the sentence. Here's an example of an unclear sentence:

> If the insulation contains a formaldehyde derivative, and if the insulation comes in contact with the wall board, and if no polyethylene barrier is used, and if no ventilation is provided for the insulation, health hazards may result.

As the *if* clauses pile up, the reader loses track of the meaning. The sentence can be improved this way:

> Health hazards may result if the following conditions exist: (a) the insulation contains a formaldehyde derivative; (b) the insulation comes in contact with the wall board; (c) no polyethylene barrier separates insulation and wallboard; and (d) no ventilation is provided for the insulation.

To separate the conditions and place them in the right-branch position requires revising and transposing, just as to avoid interrupters in the middle position, you must transpose.

8. Use the Positive Rather Than the Negative Form of a Statement

> The contradictory results did not nullify the findings of the satellite laboratory.

This sentence is hard to read because it contains three negatives. A negative is implied in *contradictory; not* and *nullify* are negatives. Researchers say that in order to comprehend negatives, the reader must transform them into their positive form. Doing that even once is extra work. Doing that more than once in a sentence and keeping all the details in mind creates difficulty for the reader. Let's put the sentence in its positive form:

> The contradictory results agreed with the findings of the satellite laboratory.

This is probably what the original means. As you can see, sentences containing multiple negatives also are often ambiguous because they contain unstated implications.

9. Choose Sentence Length and Let the Audience Help Determine Sentence Length

For some time now, reading specialists have known that a correlation exists between sentence length and readability, but even now the precise relationship is unclear. Because many other variables affect readability, we can't assume that short sentences are always easier to read than long ones are. In fact, it would be bad advice to tell writers to use short sentences always. If written effectively, long sentences can be easy to comprehend. Consider this example from *The Pittsburgh Press* (July 11, 1982). This lead sentence has 38 words:

> Call it compassion, call it common sense or "just good business," but a payroll deduction plan by United Steelworkers local 1211 to help its laid-off members is an idea that is clicking in this Beaver County community.

The *Press*, like most newspapers, is probably pitched for a reading level of high school or lower. And because this paper has been successful in attracting and keeping readers, it clearly knows how to analyze the reading level of its audience. The paper's survival depends on doing so.

In our discussion of audience analysis, readability, and sentence length, we want to follow the same principle that guides the *Press* and other successful publications of large circulation. We call this principle *laissez faire readability*. *Laissez faire* is a term taken from classical economics. It means hands off: Let the marketplace determine prices via supply and demand. Let the traffic determine conditions and set regulations. In this case the traffic is the flow of printed material in the form of newspapers, magazines, books, and other print.

We can determine the readability levels of audiences by examining the traffic flow of material that's vital to them. We examine the sentence length, as well as other factors, of the reading matter that mass audiences depend on and/or regularly pay for. If a publication is widely bought, or if it transmits the essential, everyday information of a profession or organization, it has the right readability level for that audience. That is laissez faire readability. There may be flaws in the publications that pass this marketplace test, but the writing in them is good enough to do the job.

Now, what does laissez faire readability have to do with the length of sentences in writing that you'll be doing? The answer is this. You can determine what the average sentence length (ASL) of your writing should be by looking at a generous sample of the kind of professional and nonprofessional writing that your audience reads most frequently. If you know that your audience reads *Reader's Digest*, then study that publication to determine sentence length. If an engineering periodical is standard in your profession, use its ASL as a guide to sentence length in your writing. Also prepare a list of ASLs for publications aimed at different audiences: daily newspapers, popular magazines, and professional publications. It's better to use these as standards than to follow a rule that calls for writing a lot of short sentences or that calls for a certain average number of words per sentence in all cases.

For your convenience, here is a list of average sentence lengths for a number of mass publications. The figures come from our extensive sampling, and they represent different levels of readability and different audience preferences:

Publication	ASL
New York Times	26.2
Wall Street Journal	27
Time	24.4
Newsweek	24
Popular Mechanics	21.8
Field & Stream	22.8
Science Digest	22
Various HUD Publications	22.4
Reader's Digest	20.4
Pittsburgh Press	20
Scientific American	24.9

In our view, when it comes to readability, the way a sentence is put together is probably more important than length. If you take another look at the long sentence from *The Pittsburgh Press* (see page 110), you'll notice that it is composed of short chunks of messages, little packets of information, set off from the rest by commas.

Those little packets are the result of sentence combining. The small chunks are either additions or embeddings. So the real trick to writing readable sentences, up to a point, is to give the reader information in small packets that are easy to process and remember.

In addition to ASL, you should consider another important factor: the *degree of variance (DOV)* from the ASL, that is, how short are the shortest sentences of the standard you are testing, and how long are the longest sentences? When you use ASL as a guide, you must not only observe the average length, but you should vary from the ASL just as your standard does. If the standard seems to contain about 15 percent short sentences, follow that as a guide. If it contains about 5 percent very long sentences, use that figure as your norm.

To review: Check the ASL of reading material that you know your audience reads, and use that as a guide for your ASL. Then vary the sentence length according to the degree of variance in your model. Use the DOV as a guide.

10. Use a Readable Sentence Pattern

Suppose you analyzed sentence patterns from a large body of writing in technical and other fields and discovered that one pattern was used as often as 75 percent or even 95 percent of the time. If that were the case, you could isolate that pattern and use it as a model, a kind of template, at least three-quarters of the time. In recent years investigators have, in fact, identified a basic pattern.[2] In this section we want to describe this pattern and suggest how to use it with some variation.

According to our evidence, the sentence pattern used at least 75 percent of the time is as follows:

Subject	+ Verb	+	Words to Complete the Verb	+	Optional RB Addition
1. Sonar	detects		sound waves under water,		making it a useful tool in underwater salvage operations.
2. Sonar	can be used		by military ships.		

[2]Francis Christensen was one of the first investigators to discover that professional writers don't vary their basic sentence patterns much. His findings are in "Notes Toward a New Rhetoric," *College English*, October 1963, pp. 7–18. Kenneth W. Houp and Thomas E. Pearsall affirm Christensen's conclusions and cite them as good guidelines for technical writers in *Reporting Technical Information*, 4th ed. (Encino, Calif.: Collier Macmillan, 1980), pp. 168–69. Our frequency percentages agree roughly with Christensen's.

Another 20 percent of the time, professional writers put a short addition before the subject:

Short LB Addition	+	Subject	+	Verb	+	Words to Complete the Verb
Most frequently,		sonar		is used		for navigational purposes.

About 10 percent of the time writers use long left-branch additions:

Long LB Addition	+	Subject	+	Verb	+	Completer Words
With the advent of Amtrak in 1971 and growing interest in rail transportation,		the Railway Express Building		was converted and expanded		into an Amtrak passenger station.

Let's summarize pattern frequency this way:

Pattern	*Percent Frequency*
Subject + Verb + Completer Words + (Optional) RB Additions	75
Short LB + Subject + Verb + Completer Words + (Optional) RB Additions	20
Long LB + Subject + Verb + Completer Words + (Optional) RB Additions	5

Other patterns, such as a subject delayed until after the verb, constitute a negligible percentage.

To underscore the point, consider this sentence by a readable science writer, Lewis Thomas:

> The Marine Biological Laboratory in Woods Hole is a paradigm, a human institution possessed of a life of its own, self-regenerating, touched all around by human meddle but constantly improved, embellished by it.[3]

---

[3]Lewis Thomas, *The Lives of a Cell* (New York: Bantam Books, 1968), p. 68.

Using the basic pattern, writers have much more flexibility than you might at first think. The base (Subject + Verb + Completer Words) stays the same, but you can frequently use the slot before the subject to put in a short transitional LB addition, such as *consequently, finally, however, on the contrary, in addition,* and so on. That presubject slot is perfect for *transitional words,* that is, words that serve as signposts to help the reader move from one point to the next. Here is an example of the basic pattern with a transitional phrase in the presubject, left-branch slot and with an ing-connection in the right-branch slot:

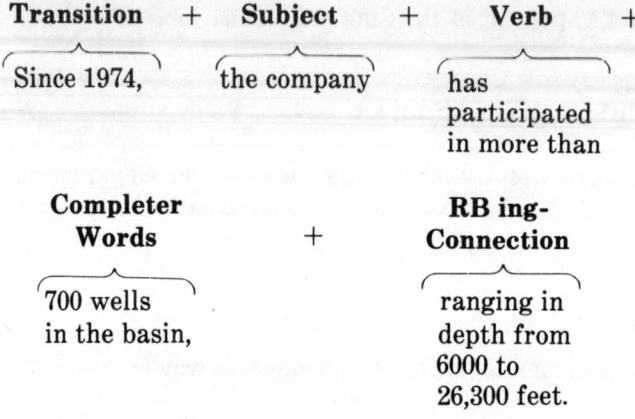

Here is another:

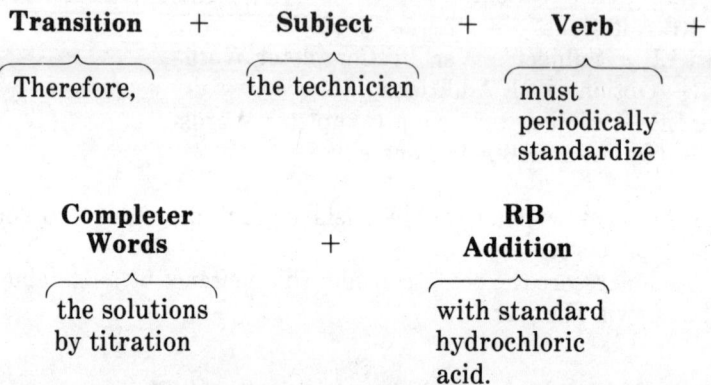

You can stack up additions in the right-branch position, as in the latter sentence, adding information to the base. You can use the right-branch position to introduce ideas that can be picked up and amplified in the next sentence. Apparently, information added to the right of the base can be more easily comprehended than information added in the left-branch or middle position. In different ways, the left-branch and right-branch positions in the sentence can be used to glue sentences together, and the right branch is a good place to add amplifying information. Let's look at this pattern schematically:

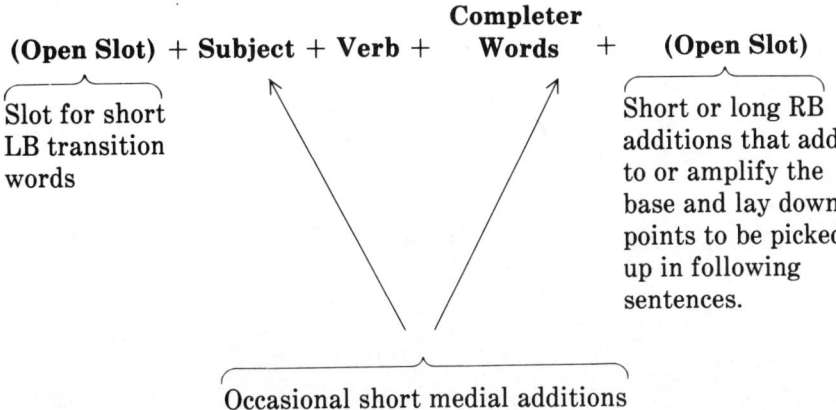

There you have it, the sentence pattern that professional writers of practical material use most often and most readers apparently find easy enough to read. As you write and edit your own material, monitor your sentences to make sure that most of them follow the (Open Slot) + Subject + Verb + Completer Words + (Open Slot) pattern. This, along with the other nine guidelines, will help you produce highly readable sentences.

EXERCISES

Revise these sentences by making the doer of the action the subject of the sentence and by using personal pronouns (see Guidelines 1 and 2).

1. The disorption efficiency must be determined over the sample range by the technicians at the installation.

2. If the measurement instructions described in the manufacturers' literature are followed, there will be small margin for error.

3. Before analysis of samples, the analytical system is allowed to equilibrate until a steady baseline is reached on the recorder.

4. Amplification of wavelengths is achieved by stimulating the quartz cell.

5. To achieve reproducible chromatographic resolution, it is critical to allow the IC column to reach equilibrium.

6. The analytical sensitivity obtained for each arsenic species under the conditions recommended in this procedure is presented in Table 1.

7. Although it is not required, it is beneficial to bypass the suppressor column and the electric-conductivity detector.

8. If conditions are employed that are different from those mentioned in this manual, blank sampling tubes should be analyzed after storage of such tubes.

9. After sample collection, the filter cassette should be firmly sealed with plugs in both the inlets and outlets.

10. The storing of an electrical charge can be achieved by use of the capacitor, which is made of plates or conducting surfaces that are insulated from each other.

• • •

Revise the following sentences by eliminating difficult words and noun clusters, and by following Guidelines 1 and 2 where applicable.

1. The particulate collection accuracy of the combined sampling and analytical method was tested at the same time as the precision by analyzing additional aerosol samples.

2. The field-sample collection efficiency of the method for organophosphates was found to be greater than 98 percent.

3. The testing of concentration levels was terminated once it was discovered that X-ray Fluorescence (XRF) analysis methods could not be utilized.

4. Particulate limit-detection instruments were found to be ineffectual for this group of PCB compounds.

5. Any ion present in the sample matrix at a high enough level can interfere with the ion chromatographic separation of the arsenicals.

6. Citizen involvement on preservation matters was actually generated by the dramatic adaptive reuse of school buildings for multiple community utilization.

7. Recycling of the preused metal inventory has had a positive effect on the visual and economic status of the immediate area.

8. Indications of future activity at the terminal are indicative of significant increases in the amount of transportation-related activity.

9. Included in these plans is the construction of a high-capacity, multiple-usage, fixed-guideway facility designed for optimum flow of pedestrian traffic.

10. Future planning favors termination of access to the communal recreation facility during inclement and off-seasonal months.

• • •

Revise the following sentences using Guidelines 5, 6, 7, and 8.

1. The drilling crew chief wanted the core samples taken every 30 feet. (Revise in two versions).

2. Molybdenum, because it is a hard, ductile, highly malleable metal able to withstand high temperatures and because it has a crystalline structure, is used in airplane and rocket parts.

3. PCB compounds, if they come into contact with the skin, if they are used in areas where inhalation is possible, if they are used in systems where particulates are suspended in the air, are subject to severe restrictions.

4. It wasn't that the CAB investigator did not want the fuselage parts put in the hangar. (Revise in two versions, according to at least two guidelines.)

5. A series of standards, varying in concentration over the range corresponding to approximately 0.1 to 3 times the OSHA standard, is prepared and analyzed under the same conditions and during the same time period as the unknown samples.

• • •

Revise the following sentences according to readability guidelines. Feel free to delete redundant information.

All transportation-sealed containers entering the yard are weighed by the yard personnel prior to out-go terminal preparation of the outbound-container manifest. The accuracy of the weight information listed on the drayman's handtag is verified by the weigh station. Cargo weight, performed upon entering the yard and prior to terminal preparation, is a factor in determining the proper billing to the shipper; thus, the terminal is insured of proper transportation-revenue collection, and an avenue for both fraudulent reduction of shipping/insurance charges—and fraudulent claims for nonexistent cargo—is blocked by carefully performing this function.

The most secure method for ensuring that only authorized cargo enters the yard is to accept only shipments with a valid terminal-booking number. A booking list is daily forwarded by the sales department to the receiving clerks. The booking number listed on the drayman's documentation is verified with the booking list to determine validity. Following verification of the outbound shipment and the physical condition of the container and chassis, the container manifest is prepared from the trucker's handtag.

If a container without proper documentation is delivered by the drayman, the terminal might assume custody of an improper container resulting in unnecessary and time-consuming legal and clerical work.

A burden on the receiving clerks is created by processing a container without documentation, and operating efficiency is slowed down. Standard receiving procedures are established by some terminal operators for containers that arrive without proper documentation to ensure the propriety of the shipment while not overburdening the clerical operations.

All the pertinent information concerning the movement and contents of a container are contained in the container manifest. The access to this document should be safeguarded by terminals by restricting its access to only a few employees and storing this document in a secure location when not in use.

Unit Nine

Nominalizations

In this section we want to talk about a special form of noun: the nominalization. *Nominalizations* are nouns formed from verbs or adjectives that have been turned into *nominals*, another word for nouns. We call this noun form special, not to praise it, but because it often leads to special writing problems—the kinds of problems that make sentences hard to read.

Recognizing Nominalizations

Verb nominalizations are frequently, but not always, long words, and you can often identify them by their endings. Notice the *ance, ion,* and *ment* endings on the nouns formed from the following verbs:

Verb	Noun Form
assist	assistance
organize	organization
improve	improvement
establish	establishment

Adjectives, too, when changed into nouns, give signals that help to identify them as nominalizations. They often end in *ness, ence,* or *ity:*

Adjective	Noun Form
worthless	worthlessness
intelligent	intelligence
complex	complexity
simple	simplicity

Wordiness and Nominalizations

Now that you have some idea of what nominalizations are, let's examine how they can create problems in writing. Before doing so, however, you should know that they don't always create problems; later in this unit, you will learn about some of the useful features of nominalizations. Nominalizations affect sentences in complex ways, and there is not enough space to describe them all here. We will concentrate on the most obvious—and most serious—problem created when a writer uses nominalizations carelessly: *Nominalizations can produce sentences that are needlessly wordy and hard to read.*

Why do nominalizations produce needlessly wordy, hard-to-read sentences? At least three interrelated factors account for this. The first probably has as much to do with writers as it does with nominalizations. Some writers, it seems, create sentences that move from one nominalization to another. This problem is the fault of the writer, who, after all, should be in control of his or her sentence. At the same time, though, it happens partly because of the nature of nominalizations: They seem to breed one another. One nominalization in a sentence tends to lead to another, the second one to still another, and so forth.

You can see this phenomenon at work even in a short sentence like the following (nominalizations appear in italics):

> The *complexity* of the algebra problem led to the *puzzlement* of the entire class.

This sentence contains two nominalizations: *complexity*, formed from the adjective *complex*, and *puzzlement*, formed from the verb *to puzzle*. In a truly efficient, streamlined form, the sentence would read:

> The complex algebra problem puzzled the entire class.

This version returns both nominalizations to their original forms. *Complexity* becomes the adjective *complex*, and *puzzlement* appears as the verb *puzzled*. This is one—and perhaps the ideal—revision of the original sentence, but the writer has options, too. For example:

> The complexity of the algebra problem puzzled the entire class.

In this case, the writer does not allow the first nominalization, *complexity*, to lead to a second, *puzzlement*, but rather uses *puzzled* as the verb.

Although this sentence is less efficient than the earlier revision, it conveys its message more efficiently and clearly than the original.

Grammarians do not know why some writers let one nominalization lead to another. They suspect the habit develops very early, perhaps in college. Many academic writers use a highly nominalized style. Their students, after reading large quantities of nominalized writing, eventually absorb that style, more or less subconsciously, and then repeat it in their own writing. Nevertheless, this doesn't have to happen. You can control nominalizations, and you can do so even while falling short of producing the ideal sentence. As the previous example suggests, you don't have to let one nominalization lead to another. If you learn to recognize nominalizations and know when you are using them, then you won't write sentences such as the following (nominalizations are in italics):

1. A *reduction* in the company's work force was the *result* of the *lack* of a favorable *response* by too many customers to its new *products*.

2. If *cooperation* from our workers is forthcoming, the *achievement* of a *reduction* and, finally, the *elimination* of assembly-line *slowdowns* will result.

Later we will examine these sentences closely and present a method for revising them, but first we want to point out another reason that nominalizations produce wordy sentences.

Prepositions and Nominalizations

Not only do nominalizations spawn other nominalizations, but they also spawn prepositional phrases. Look closely at sentences 1 and 2 and notice how often prepositional phrases follow the nominalizations. Nearly every nominalization demands an accompanying prepositional phrase:

reduction *in the company's work force*

lack *of a favorable response*

elimination *of assembly-line slowdowns*

Prepositional phrases may follow any noun, but with nominalizations, prepositional phrases too often have to follow. This situation results from the abstract nature of nominalizations. Take a look at the nominalizations in examples 1 and 2 and you will notice that words such as *reduction, response, cooperation,* and *achievement* are all abstract nouns that name concepts rather than living things or objects in the physical world. There are exceptions, but often nominalizations name things that exist only in our minds, not in the world known to our senses. They seldom name things we can see or touch; rather, they name *concepts* that are familiar to us.

Because nominalizations describe abstractions, writers have a tendency to add words to make the nominalizations clear. Unless a nominalization has been defined in preceding sentences, there is a tendency to follow it with a defining prepositional phrase. Example 1 demonstrates how these definitions can accumulate because of the nominalizations (prepositional phrases are italicized):

1. A reduction *in the company's work force* was the result *of the lack of a favorable response by too many customers to its new products.*

In this sentence, the writer wants to tell the reader why a company reduced its work force. Notice how prepositional phrases appear in succession when the writer tries to explain why:

was the result *of the lack/ of a favorable response/ by too many customers/ to its new products.*

Notice also how the reader is left with an unanswered question after each prepositional phrase:

was a result	(of what?)
of the lack	(of what?)
of a favorable response	(by whom? to what?)
by too many customers	(to what?)
to its new products	

Prepositional phrases can make sentences difficult to read. So many prepositional phrases in a row force the reader to store little nibbles of information in his or her mind as the sentence slowly reveals what those little nibbles all add up to. This process slows the reading pace and sometimes makes the reader glance back over the sentence to check its meaning.

In fact, a good way to identify heavily nominalized sentences is to look for those cluttered with prepositional phrases. Then, take a second look to see if they are also cluttered with abstract nouns several syllables long.

Nominalizations as Subjects

By this time we hope we have convinced you that nominalizations often produce sentences that are too wordy and harder to read than they should be. There is a reason for this: *Writers produce heavily nominalized sentences because they pay too little attention to choosing the right sentence subject.*

Nominalizations usually work poorly as sentence subjects. They confuse the actor-action relationship that normally exists between subject and verb. This is the chief cause of the wordy, hard-to-read sentences associated with nominalizations.

As you will recall from an earlier discussion, the subjects capable of producing the most efficient sentences are subjects that *perform or cause the action* mentioned by the verb. Nominalizations rarely name things that are capable of doing either. Notice the verb produced by the nominalized subject *reaction* in this short sentence (subject and verb are in italics):

3. The machinist's *reaction* to the warning buzzer *was* immediate.

As short as it is, this sentence shows what often happens when a writer chooses a nominalization for a subject. It leads to a verb that expresses no action at all. In this case, *reaction* produces *was*, which merely connects the subject *reaction* to the adjective *immediate*.

This short sentence is not particularly inefficient, but a more efficient version would use *reaction* as the verb and *machinist* as the subject. A machinist is capable of reacting to the buzzer:

The *machinist reacted* immediately to the warning buzzer.

In the original sentence, which simply intends to say that the reaction came quickly, the verb options almost disappear once the writer chooses *reaction* as the subject. *Was* is nearly inevitable.

Nominalizations and Verbs

Nominalized subjects present relatively few verb options, and the verbs that can accompany them are frequently as imprecise in meaning as the nominalizations themselves. On the other hand, subjects capable of acting can produce long lists of verbs that express specific actions. To test this, let's experiment with the subject *machinist* from example 3.

A noun such as *machinist*, which names a living being, has the power to generate many highly specific verbs. A machinist, for example, can *hear* the buzzer; he can *reach out*, *press* the power button, and *turn* the machine off. Or he can *flip* or *pull* or *slide* or *crank* the safety shield down over the lathe, automatically silencing the buzzer. He can also *breathe* a sigh of relief afterward, or *mumble* or *complain* about hazard-warning buzzers and safety regulations.

By contrast, *reaction*, the original subject of example 3, can generate only a few verbs. To test this, try to draw up a list of verbs that will work with *reaction*:

The reaction *was* one of fear.

The reaction *was noticed* or *was seen* or *was observed* by everyone.

The reaction *led* to changes in . . .

These verbs are not nearly so action-oriented or descriptive as the ones that would be used when *machinist* is the subject. Can you add many more? There are some other possibilities, but most are fairly static.

If you team *reaction* with a verb that leads to certain other nouns or pronouns, you can extend the verb list. For example:

The reaction *stunned* us.

Or you might write:

His reaction *shocked* or *amazed* or *frightened* or *reassured* us.

Now the list becomes longer, the verbs more specific. Most any nominalization that will take a possessive pronoun can work like this: *his inadequacy, her observation, their achievement,* for example. Nevertheless, this extra verb range is possible only because a living being, capable of acting or reacting, is named later in the sentence. Even so, the verb list produced by using nominalizations in this way will not be long when compared with the verb-producing abilities of nouns and pronouns like *machinist, engineer, team, crew, he, she, they.*

Experiment with other nominalizations. You will find that they are somewhat limited in their capacity to generate a list of verbs, especially action verbs.

Revising Nominalized Sentences

Using nominalizations as subjects leads to reduced verb options, but it also leads to fewer options for structuring the rest of the sentence. A writer, having chosen a nominalization as a subject, risks losing control of the entire sentence.

Most any heavily nominalized sentence will demonstrate this, but let's look at example 1 again (nominalizations are in italics):

1. A *reduction* in the company's work force was the *result* of the *lack* of a favorable *response* by too many customers to its new *products.*

Here, the nominalized subject, *reduction,* produced the verb *was.* With this subject and this verb, what options remain for continuing the sentence? Not very many. If you choose to tinker with the verb, you might consider such options as *came about* or *became inevitable:*

A reduction in the company's work force *came about because* of the lack of a favorable response by too many customers to its new products.

A reduction in the company's work force *became inevitable because* of the lack of a favorable response by too many customers to its new products.

These verbs allow you to drop the nominalization *result,* but the sentence has not changed much. If you continue to tinker, you might notice that *favorable* could be changed to *unfavorable,* allowing you to drop the prepositional phrase *of the lack:*

A reduction in the company's work force became inevitable because of an unfavorable response by too many customers to its new products.

With this last effort, you have eliminated two nominalizations, *result* and *lack*, and the prepositional phrase *of the lack*. The sentence is now slightly less wordy, somewhat more readable. But it isn't dramatically improved.

Tinkering won't help much as long as you ignore the nominalized subject. To improve this sentence, you must find a subject capable of action, and *reduction* lacks that capability. Fortunately, a noun that names someone or something capable of acting frequently appears somewhere in a heavily nominalized sentence. Here, the noun is *company*. Now you have an actor as subject. If you change the former subject, *reduction*, to its verb form, you will have a verb that rather specifically names the action taken by the subject:

> The company reduced its work force...

With this start, you can now produce an efficient sentence. Look at the remainder of the original sentence. Notice that you have the materials there for a similar subject-verb grouping:

> ... too many customers responded unfavorably.

Linking these two clauses to form a strong sentence shouldn't be difficult:

> The company reduced its work force because too many customers responded unfavorably to its new products.

This is a less wordy, easier to read sentence than the original. It could, of course, be streamlined further:

> The company reduced its work force because not enough customers bought its new products.

Or:

> The company reduced its work force because it could not sell enough of its new products.

Or you might reverse the two clauses:

Because it could not sell enough of its new products, the company reduced its work force.

You can now tinker productively because you have coupled subjects capable of acting with verbs expressing specific actions.

This, then, is the way to tinker with a heavily nominalized sentence. Find a subject that can act or cause action. Remember, a potential subject may lie hidden somewhere in the sentence. If the original subject is a nominalized verb, it can serve as the verb for the sentence. In fact, if the original subject is a nominalized verb, begin your tinkering there. Once you think of it as a verb, the appropriate subject will likely come to mind.

Now, let's tinker with example 2 (nominalizations are in italics):

2. If *cooperation* from our workers is forthcoming, the *achievement* of a *reduction* and, finally, the *elimination* of assembly-line *slowdowns* will result.

The subject of the first clause, *cooperation*, is a nominalized verb. If you return it to its verb form, *cooperate*, you can then search for a subject. Who does or will do the cooperating? Obviously, the workers. This gives you a subject and verb for the subordinate clause:

If our *workers cooperate* ...

Before tinkering further, you might note that this is a cause-and-effect sentence. The opening subordinate clause states a conditional cause (if the workers cooperate), and this condition, if fulfilled, will have an effect (reduction and elimination of slowdowns). But you don't want to retain the string of nominalizations used to explain the effect. This would not improve much on the original:

If our workers cooperate, the *achievement* of a *reduction* and, finally, the *elimination* of assembly-line slowdowns will result.

Since *reduction* and *elimination* are nominalized verbs, you can return them to their verb forms, *reduce* and *eliminate*. These express more precisely defined actions than the original verb *will result.*

To complete the revision, you need a subject. Since the first clause mentions *our workers*, you can use the pronoun *we* as an actor subject:

If our workers cooperate, *we* can reduce and, finally, eliminate assembly-line slowdowns.

Lining up the subject and verb of each clause into an actor-action sequence allows you to drop the nominalizations and prepositional phrases. The result is a more efficient, easier-to-read sentence. Technically, the revised sentence does contain one nominalization—*slowdowns*. This is a nominalized verb, but it works. You would have a hard time finding a noun to replace it.

All the sentences discussed so far have had an actor appearing somewhere in the sentence. Occasionally, you will find a sentence with a nominalized subject but with no actor buried in the sentence:

4. An expansion of the work force occurred.
5. There was a renovation made to the building.

To revise such sentences, find the actor or the person or the thing that caused the action. In example 4, ask yourself, "Who did the expanding? Who was responsible for causing the expansion?" The *work force* didn't do or cause the expanding. The expanding *affected* the work force by increasing its size. Finding the actor should produce a sentence something like this:

The construction company expanded its work force.

Now the subject and verb show an actor-action relationship. Example 5, in revised form, might look like this:

The owner renovated the building.

Be a bit cautious when revising actorless sentences. You will have to determine who or what has performed or caused the action. Don't be misled by other nouns that appear in the sentence. Using one of them will lead to passive constructions like these:

4. The work force was expanded.
5. The building was renovated.

These two sentences seem harmless enough. They are certainly easy to read. But passive constructions, like nominalizations, often lead to wordy, inefficient sentences.

Useful Nominalizations

While nominalizations do require careful use, you could not stop using them even if you wanted to. Some nominalizations have become so well accepted as nouns that we must use them in certain writing situations. Many of those made from verbs and ending in *ing* fall into this category: *fishing, jogging, talking*, for example. These are called *gerunds* when used as nouns, and many of them are much more precise in meaning than most nominalizations. This is especially true of those that refer to very specific actions. Nevertheless, they do not make particularly good subjects because they work well with only a few verbs.

Other nominalizations are so common in specific contexts that you cannot avoid them. At colleges and universities, professors give *assignments* and students write *reports*. In hospitals, surgeons perform *operations*. On the job, we work for *pay* and complain about the *deductions*. At *meetings*, we vote on *motions*, and so on.

Sometimes, nominalized subjects can make your writing more coherent. They can serve as transitions by renaming, in a different way, a topic mentioned in previous sentences. The following italicized nominalizations work in this way:

> The engineers had pictured an unearthly, complex gadget with enough wires and tubes protruding from it to fill a large room. Then last week they saw the new engine for the first time. Its *simplicity* astounded them.
>
> Nitinol, a metal made from nickel and titanium, responds in odd ways to slight differences in temperature. For example, if a length of nitinol wire is bent while cold, it will snap back into its original shape when heated. This *characteristic* excites those who think of nitinol as a new energy source.

A Final Word

The more you write, and especially, the more you tinker with sentences, the more you will learn about when to use nominalizations and when to avoid them. You now know what nominalizations are, and you have some idea how using them carelessly can make sentences wordy and hard to read. You know how to identify such sentences when proofreading, and you know where to begin your tinkering when you set out to revise them.

EXERCISES

In the sentences that follow, make the actor-action relationships as clear and specific as possible by eliminating as many nominalizations as you can. Remember, though, that you don't have to get rid of every single nominalization. Let your ear tell you when you have constructed an efficient sentence. If a sentence doesn't name an actor, use your imagination to supply one in your revision.

Wherever possible, use the combining strategies covered earlier in the book. To assist you, we have italicized the key nominalizations in every other sentence.

1. *Improvements* in quality control have been effected in our three plants in the Northeast.

2. A provision for easy maintenance is the sliding cover over the drive belt.

3. Among our *plans* for next year is more *efficiency* in the *utilization* of office space.

4. The replacement of wheel bearings is a job that can be handled by the least experienced mechanic.

5. *Investments* in microchip research will be made in 1984.

6. If an identification of minor flaws in our metal casings can be made before shipment, there will be fewer rejections of units by our customers.

7. There were continuous *vibrations* in the pulley shaft.

8. We made a careful consideration of costs before reaching the decision to sell our smelting plant.

9. The *conveyance* and *implementation* of programming instructions from the main memory of a computer is performed by the central processing unit.

10. We finally reached a decision calling for the substitution of coal for oil in our generating units.

Unit Ten

Paragraph Cohesion

The process of writing is one of binding sentences together into longer units called paragraphs, which, in turn, become even larger units, such as reports, articles, chapters, and books. *Cohesion* refers to how well the various units of writing are bound together.

In this unit you will see how certain words and phrases contain the mortar, the binder, that makes paragraphs cohesive. You will also see how the arrangement of information within individual sentences can add to paragraph cohesion.

You are already familiar with some of these methods for binding paragraphs together. Others you are probably aware of intuitively, but you may not have thought about them before. One reason you can make these work is that your readers are also intuitively aware of them.

As you add one sentence to another to produce a paragraph, each sentence depends in some way on the sentences that have gone before it. In a sense, the experienced writer keeps looking back at the earlier sentences because they determine which binders he or she should use. Except in proofreading and editing, the writer, of course, does this intuitively.

When proofreading, you can easily check to see whether you looked back intuitively while producing your rough draft. If you did, your paragraphs will contain words that look back to words in earlier sentences. Notice how the italicized words in the following paragraph look back to the word in the box:

All boxed(fasteners) are designed for one purpose—to attach components together securely. *Some* are used advantageously in woodworking.

Others have special applications for *fastening* metal parts. *Still others* are used to accelerate *fastening* and *unfastening* panels. Let's examine the distinctions between various *kinds*.[1]

The italicized words keep *fasteners*, the topic of the paragraph, in front of the reader. They do so by binding the topic into the whole paragraph. By using *some, others,* and *still others,* the writer achieves the effect of using the noun *fasteners* as the subject in each of the first four sentences. And what the writer gains from this is worth examining. He is able to keep the reader's attention focused on the topic rather than on the writing. He has kept the reader from noticing the writing.

Obviously, using *fasteners* as the subject in all four sentences would be awkward and distracting to most readers, as this example shows:

> All *fasteners* are designed for one purpose—to attach components together securely. *Fasteners* are used advantageously in woodworking. *Fasteners* have special applications for fastening metal parts. *Fasteners* are used to accelerate fastening and unfastening panels.

The writer of the original paragraph avoids distracting the reader by using words like *some* and *others*. He keeps the topic in front of the reader and, at the same time, avoids repeating the original noun over and over.

However, since words such as *some* and *others* take their meaning from an earlier noun, the writer must not drift too far from *fastener* before introducing them. Notice what could happen if the paragraph about fasteners were allowed to drift:

> All fasteners are designed for one purpose—to attach components together securely. People who work in the building trades use fasteners often, since their jobs involve attaching parts onto other parts. *Some* are used advantageously in woodwork. *Others* have special applications for fastening metal parts. *Still others* are used to accelerate fastening and unfastening panels.

---

[1] *Rate Training Manual,* NAVPERS 10085-B (Washington, D.C.: U.S. Government Printing Office, 1971), p. 112.

The second sentence has introduced a new topic—people who use fasteners. In doing so, it also introduces the pronouns *who* and *their*. By the time the word *some* appears, the reader can no longer be certain whether it refers to *people* or to *fasteners*. The paragraph lacks cohesion because it fails to keep a clear topic in front of the reader. Being aware that binding words are expected intuitively by a reader will help prevent drifting away from the topic. This awareness will make you think about binders from the first sentence onward.

Variations, Repetitions, and Substitutes

By shifting quickly to *some, others,* and *still others,* the writer of the original paragraph gains another advantage. Because the word *fasteners* doesn't appear four times in subject slots, the writer is free to use *variations* of the word in some of the sentences:

> Others have special applications for *fastening* metal parts. Still others are used to accelerate *fastening* and *unfastening* panels.

As part of the accumulation of binders, these variations keep the topic close to the reader. *Repetitions* are another important binder. When we discouraged the repetition of *fasteners*, we weren't suggesting that words or phrases should not be repeated. Repeating a noun like *fasteners* at the beginning of four consecutive—and short—sentences can have a distracting rather than a binding effect, but nouns are often repeated—or repeated in some variation—in order to bind in the topic. Notice how repetition and variation are used in this example:

> ⎡Nails⎤ achieve their fastening or ⎡holding power⎤ when they displace ⎡wood fibers⎤ from their ⎡original position.⎤ The pressure exerted against the *nail* by *these fibers*, as *they* try to spring back to their *original position*, provides the *holding power*.[2]

See how the boxed terms in the first sentence are repeated or varied in the second sentence:

---

[2]*Rate Training Manual*, NAVPERS 10085-B (Washington, D.C.: U.S. Government Printing Office, 1971), p. 112.

Sentence 1	Sentence 2
Nails	nail
holding power	holding power
wood fibers	these fibers, they
original position	original position

Nails is repeated in its singular form. *Holding power* and *original position* are repeated exactly, while *wood fibers* is first repeated as a slight variation, *these fibers*, and again as the pronoun *they*. This writer began looking back and binding his meaning in very quickly.

Variations have a wider range than that shown by this paragraph. They can be used to look back to units longer than one or two words. In the following sentence, notice how the variation *any service station mechanic* looks back to the embedded clause *someone who knows something about engines:*

If you can't find the PCV valve, have *someone who knows something about engines* check to see if it might be clogged. *Any service station mechanic* can locate and test a PCV valve in a matter of minutes.

Substitutes are another type of binder. They are usually pronouns; as such, they stand in for nouns. Personal pronouns such as *he, she, they,* and *them* are probably the pronoun substitutes most familiar to you. However, pronouns like *some* and *others* form a large group of substitutes. Because this group refers loosely to measurements, the pronouns in it are particularly important in making comparisons, explaining functions, and describing alternative methods of doing something. The following pronouns belong to this group:

several	all
another	most
both	many
each	few

These pronouns often appear as substitutes along with more precise numbering binders, such as *one, two,* and so forth; *first, second,* and so on; *dozens, hundreds,* and the like.

Although these more precise numbers are usually thought of as nouns, they, too, receive much use as substitutes. Notice how *one* and *second* look back to the noun *patent* in the example:

His first *patent* was merely a clever *one*. The *second* was a stroke of genius.

Notice how all the substitutes in the following example form a network of binders, with each one helping to bind in the topic *devices for mounting and securing car batteries:*

> Over the years, automotive engineers have devised literally *dozens* of devices for mounting and securing car batteries. *Most* of these combined simple but clever ideas with inexpensive materials. *Some* of them, to the dismay of owners and mechanics, reflected the complex thinking of theoretical geniuses who were practical dummkopfs. A *few* even used expensive materials. Only *one* or *two* have remained unchanged to this day. *All* of them, however, usually managed to do what they were intended to do: Hold the battery securely in defiance of the laws of gravity and motion.

Notice that several of the substitutes are followed by additional binding phrases: *Most of these, some of them, all of them.* This is a common binding arrangement when numbering words act as substitutes. The combination of words enables the reader to maintain a clear and precise relationship between the topic and the binder.

Numbers also act as adjectives and combine with nouns to produce variations:

> If you have doubts about the software programs suggested by *one consulting firm*, then ask for an opinion from a *second group of experts.*

The *demonstrative pronoun binders—this, that, these,* and *those—* work in the same way as numbers, and they are just as important to paragraph cohesion. They can stand alone as substitutes:

> Our Plastic Container Division increased its sales by 40 percent. *This* is close to the increase we had predicted as the year began.

In this example, *this* reaches back and substitutes for the 40 percent sales increase. Demonstratives also work as adjectives to support variations:

> In 1981 our Plastic Container Division increased sales by 20 percent and revenues by 32 percent, but profits decreased by 12 percent. *This decrease*, following rising sales and revenues, shows how inflation can exert considerable force on a balance sheet.

Demonstratives also are important binders because we use them so often in conversation to direct the listener's attention toward something specific. Occasionally we accompany the word with a pointing gesture and say something like, *"That* car over there." On other occasions we direct attention by a display of some sort. We might, for example, hold a pencil between thumb and forefinger, switch it back and forth, and say, *"This* pencil." Thus, all four demonstratives carry strong associations from speech to writing. We expect them to direct our attention to something around us. In writing they look back to something that has already been named.

Location Markers

Here, there, now, and *then* are also important binders. They are similar to *this, that, these,* and *those* in that they point. But their pointing is to *locations in space* and *locations in time.*

Here and *there* are space locators:

> Lift the distributor cap. *Here* you will find the rotor. Find the timing mark next to the V-belt pulley. Aim the timing light *there.*

Here and *there* also help smooth the way when you are directing the reader's attention to something in your own writing. If you are ready to present examples to illustrate what you have just explained, you might say, *"Here* are two examples." *There* is used in the same way, except that it points to things further away in your writing, as this example shows:

> For additional details on how to use the miterless cut on corner moldings, see Appendix I. *There* you will find . . .

Because they locate something in space, *here* and *there* are particularly important binders for describing mechanisms, comparing and contrasting two or more items or mechanisms, writing instructions, referring to blueprints or other drawings, or pointing to graphs, charts, drawings, illustrations.

The time locators, *now* and *then,* are used frequently in the writing of instructions and in describing or explaining processes—two important kinds of writing in the world of work. They work not by looking back to nouns but by reminding the reader of movements through a series of steps or a sequence of time, as this example shows:

> *Now* let the sediment settle to the bottom. *Then,* after the liquid is clear, use a siphon to draw off approximately one pint.

Signpost Binders

Signpost binders are somewhat more noticeable than the groups of binders discussed previously. They are like the highway signs that help you make the right connections at interchanges and intersections. Signpost binders connect sentences and parts of sentences. They give the reader a sense of direction in relation to the topic of a paragraph.

There are several groups of signposts. In the following pages we will build a paragraph with signposts in it from which we'll draw the groups. That way you'll see them at work as you read. Here are the first two sentences of the paragraph:

> (1) We have arranged your training schedule so that you will spend most of your time working with one of our managers. (2) *However, on some occasions* we will assign you to work with a new trainee who has just started in the program.

See how the first signpost, *however,* directs the reader forward by suggesting that sentence 2 will somehow modify what was said in sentence 1. Without the *however,* a reader would expect more talk in sentence 2 about spending time with managers. The binder *however* tips the reader off to expect something *contrary* to what he or she has just been told.

Now let's add another sentence:

> (1) We have arranged your training schedule so that you will spend most of your time working with one of our managers. (2) *However, on some occasions* we will assign you to work with a new trainee who has just started in the program. (3) *At other times* we may ask you to spend several days with a trainee who is more advanced in the program than you are.

In this example, you see *time signposts* at work. The binders *on some occasions* and *at other times* tell the reader that the contrast signaled by *however* comes in two parts. Each signpost also reaches back to sentence 1 by pointing out a situation contrary to the situation mentioned in sentence 1.

All three sentences deal with events taking place in time. The signpost *however* tells the reader that these events will differ from the first one mentioned. Here is a diagram:

Sentence 1   ... spend most of your time with a manager.

          However,

Sentence 2   ... *on some occasions* there will be exceptions to spending time with a manager.

Sentence 3   *At other times* there will be another exception to spending time with a manager.

Notice that *however* works with two sentences and sets the stage for two exceptions. *On some occasions* and *at other times* introduce those exceptions.

By now you have probably thought of other signpost binders that can give signals similar to *however*. Some of these are:

but	on the other hand	instead
yet	on the contrary	instead of
though	in spite of	rather
even though	despite	rather than
although	even so	

Remember, these signposts contrast one bit of information with other bits of information. They tell the reader that what follows will say something contrary to what was just said.

The following are some other signposts:

next	later on	in the future
after	previously	soon
afterward	eventually	as soon as
before	meanwhile	at this time
beforehand	finally	until
before long	later	since then/since that time

Now let's continue to build our paragraph and to identify other kinds of signpost binders:

(1) We have arranged your training schedule so that you will spend most of your time working with one of our managers. (2) *However, on some occasions* we will assign you to work with a new trainee who has just started in the program. (3) *At other times* we may ask you to spend several days with a trainee who is more advanced in the program than you are. (4) Both experiences are designed to enable you *not only* to learn from other trainees *but also* to teach other trainees during your apprenticeship. (5) We believe this alternating reinforces learning *and* hastens the training process. (6) We, *therefore*, expect you ...

Not only and *but also* in sentence 4 and *and* in sentence 5 belong to the same signpost group. They are *addition signposts*. *Not only* tells the reader that information is being added in two pieces. *But also* introduces the second piece of information—*not only* metal stampings *but also* metal rods.

The word *and* is so common that it is barely noticed, but it is another addition signpost. It is a sign to the reader that similar information will follow. It connects similar and roughly equal things—like nuts *and* bolts. At the same time it signals that something is being added.

Here are some other addition signposts:

also	in addition	further
and also	additionally	furthermore
either ... or	likewise	moreover
neither ... nor	similarly	besides this or that

Unfinished sentence 6 contains the signpost *therefore:*

(5) We believe this alternating reinforces learning and hastens the training process. (6) We, *therefore*, expect you ...

You could finish the sentence like this:

We, *therefore*, expect you to be enthusiastic when working with a manager, when learning from a more experienced trainee, or when teaching a less experienced one.

Or like this:

We, *therefore*, expect you to share what you learn as well as to apply what you learn.

In whatever way you decide to finish the sentence, the word *therefore* tells the reader that the sentence will draw a conclusion based on the belief that learning and teaching hasten training.

Therefore connects by reaching back to the previous sentence and, at the same time, by pointing ahead to a *conclusion* or *result* based on the previous sentence. Most signpost binders in the *therefore* group work in this same way:

Consequently, we expect you ...

So, we expect you ...

Because of this, we expect you ...

Thus, we expect you ...

For this reason, we expect you ...

With this belief *in mind*, we expect you ...

As a result of this belief, we expect you ...

It follows that we expect you ...

Accordingly, we expect you ...

Signposts binders work hand in hand with the other binders discussed earlier in this section. They are more noticeable than those mentioned earlier, but they work just as hard at binding sentences together and in guiding the reader through your paragraphs. Both groups of binders are easy to use, and you will want to use them to make your paragraphs more cohesive.

Old and New Information

In your final review of what you have written, quickly check your sentences to see how you have arranged the information in them. Generally speaking, sentences contain two kinds of information—old and new. Both terms refer only to the subject you're writing about. They have nothing to do with what the reader might already have in mind when he or she sits down to read what you've written.

Old information is the information that you have already mentioned to the reader. New information is the information that you haven't yet mentioned. When you explain something, the new information is what carries the explanation forward. The old information is what you have already explained.

Although you don't need special skills to arrange old and new information in your sentences, the following principle will help: *Paragraphs are easier to read if the old information appears in the first part of your sentence and the new information comes toward the end of the sentence.* For example, assume that the following sentence opens a paragraph on radial arm saws:

(1) Listen carefully as you flip the power switch and start the saw.

Since this is the first sentence, let's agree that all the information is new. Thus, any of the information here may be repeated as old information in the second sentence. Remember, we'll expect the old information to appear first, and so the second sentence might follow the first in this manner:

(1) Listen carefully as you flip the power switch and start the saw. (2) As the *saw* blade revolves, the noise it makes at first will be a light, whirring sound.

Sentence 1 talks about a saw. Sentence 2 begins with "As the *saw* blade revolves." You already recognize *saw* or *saw blade* (since everyone knows that saws have blades) as a binder. Because *saw* is mentioned in sentence 1, it represents old information. The new information in sentence 2 describes the noise the saw will make as the blade begins to turn. We can predict that the first part of the next sentence will reach back and present something about the noise or sound:

(1) Listen carefully after you flip the power switch and start the saw. (2) As the saw blade revolves, the noise it makes at first will be a light, whirring sound. (3) *This whirring* will quickly increase in intensity until it becomes a steady, vibration-free hum.

Again the new information from sentence 2, the whirring sound, becomes the old information in sentence 3. The new information tells us that the whirring, which you already know about, will become a hum, setting the stage for sentence 4:

(1) Listen carefully after you flip the power switch and start the saw. (2) As the saw blade revolves, the noise it makes at first will be a light, whirring sound. (3) This whirring will quickly increase in intensity until it becomes a steady, vibration-free hum. (4) *The humming sound* tells you that the blade has reached its maximum cutting speed.

As you check on the arrangement of old and new information in your sentences, don't be overly fussy when deciding where the first part of a sentence ends and the second part begins. Just think of subjects and near-by modifiers as the first part and of predicates as the second part. This works well when the subject is introduced quickly and the predicate follows without interruption:

(1) *Socket wrenches* come in several forms. (2) *One common type* has a ratchet built into its head.

The italicized subjects here form a good breaking point between first part and second part. If these sentences opened a paragraph, you could tell quickly that the subject of the second sentence contained old information: *one common type* (of socket wrench). You could also tell just as easily that this old information had been drawn from the first part of sentence 1.

In some complexly structured sentences, using subjects and predicates as guides doesn't work so well. Look again at sentences 1 and 2 from the paragraph on radial arm saws:

(1) Listen carefully after you flip the power switch and start the *saw*. (2) *As the saw blade* revolves, the noise it makes at first will be a light, whirring sound.

Sentence 2 begins with a subordinate addition: *As the saw blade revolves*. This contains the old information even though the main clause that follows contains the subject of the sentence, which is *noise*.

The old information in a sentence may come from either part of the preceding sentence. However, your sentences should be fairly consistent. If your second sentence draws old information from the first part of the preceding sentence, then make most of your sentences do likewise. Notice how each sentence in the following paragraph repeats—with variation—the subject of the opening sentence:

(1) *Socket wrenches* come in several forms. (2) *One common type* has a ratchet built into its head. (3) *Several other types*, however, have no ratchets. (4) *One of these*, the speed handle wrench, is used for spinning the nut off or on but not for loosening or tightening. (5) *Another ratchetless socket wrench* is the sliding T-bar.

The pattern here is consistent. Each succeeding sentence draws its old information from the first part of the preceding sentence. When this is the case, the old information will usually be a repetition, a variation, or a substitution of a previous topic. These were called binders earlier in this section, and they are still binders because the whole concept of placing old information first is a larger-scale method of binding to achieve paragraph cohesion.

And the pattern in the preceding paragraph wouldn't have to be consistent. The last sentence, for example, could read:

(5) *Loosening and tightening* require more torque than the speed handle can exert.

In this case, the old information—loosening and tightening—is drawn from the second part of the preceding sentence and could be used to start the new sentence, depending on how much information the writer wanted to give about the speed handle wrench. Nevertheless, the sentence would still place old information first, new information second.

It is best to try to be fairly consistent in your patterns. What is important, though, is to strive for consistency in old and new information. Reversing the patterns occasionally doesn't matter so much. But, as often as possible, do place the old information first in the sentence, the new information last—old information in the left part of the sentence, new information to the right.

Begin to apply the old-new concept after you have a rough draft. Then you can tinker with more ease, and you can start by checking old-new in the paragraphs that you're not happy with. A failure to put old information first may turn out to be the reason you weren't happy with the paragraph in the first place.

Nevertheless, these reversals will occur occasionally. Finding one means asking yourself why it is there, deciding how easily it can be reworked or whether you want to leave it as it is. But finding many reversals usually means poor paragraph cohesion and hard work for the reader.

EXERCISES

1. In the following paragraph, we have put boxes around several words or phrases that produce binders. Circle the binders that look back to each boxed word and decide whether the binders are repetitions, variations, or substitutions. Then look for signpost binders.

   boxed Four-cylinder engines are increasing in popularity in America. Because these smaller engines operate at more revolutions per mile, the boxed plugs fire more times per mile and spark-plug wear occurs in them. Additionally, a malfunctioning spark plug is more apparent in smaller engines. Four-cylinder engines thus offer outstanding opportunities for spark-plug sales. Market research shows that, on the average, boxed owners of vehicles with smaller engines replace spark plugs at shorter mileage intervals than do owners of 8-cylinder engines.[3]

2. In the following paragraph, first look for repetitions. Make a list of them. Does the number surprise you? What other binders do you find in this paragraph?

   A certain amount of the force applied to a tackle is lost through friction. Friction will develop in a tackle by the lines rubbing against each other, or against the shell of a block. Friction also is caused by the passing of the line over the sheaves. An adequate allowance for the loss due to friction must be added to the weight being lifted in determining the power required to lift a given load. Roughly, 10 percent of the load must be allowed for each sheave in the tackle.[4]

3. What binders dominate, or are the key ones, in the two following paragraphs? Are these the binders you would expect when a writer describes or explains a process? Why? What other binders do you find?

   This is a self-instructional program and might be a little different from most books that you're used to. Here's how it works. Each page is called a frame. In a frame, we will ask you to read and absorb a fairly small amount of information. Then, to make sure you've

---

[3]Champion Spark Plug Company, *1980 Annual Report*, p. 7.

[4]*Builder 3 & 2* (Washington, D.C.: Bureau of Naval Personnel, 1970), p. 68.

learned and remember it, we'll stop and ask you to answer some questions. When you have done this, we will tell you right away whether or not you were right. If you gave any wrong answers be sure to review the material before continuing. You can work at your own pace and may stop any time for a break. Now, find a comfortable, well-lighted, and quiet place to work. Then turn the page and begin.[5]

• • •

Another use of the center punch is to make corresponding marks on two pieces of an assembly to permit reassembling in the original positions. Before taking a mechanism apart, make a pair of center punchmarks in one or more places to help in reassembly. To do this, select places, staggered as shown in figure 1–47, where matching pieces are joined. First clean the places selected. Then scribe a line across the joint and center punch the line on both sides of the joint, with single and double marks as shown, to eliminate possible errors. In reassembly, refer first to the sets of punchmarks to determine the approximate position of the parts. Then line up the scribed lines to determine the exact position.[6]

4. First, locate the signpost binders in the following passage. Then identify the other kinds of binders.

Our aggressive new business effort has begun to pay off. In particular, we have been highly successful in developing new customers west of the Mississippi. We will augment this effort in fiscal 1981 with an entirely new marketing program targeted to the immediate business needs of our customers and prospects.

Looking farther ahead, we see positive factors that far outweigh the effects of some negative trends. On the negative side, new car sales are off. But this is offset by the fact that older cars are being scrapped at a lower rate, leaving a net increase in total vehicles on the road.

The key to parts replacement demand is the accumulated miles driven per vehicle. Even though the total miles driven has declined over last year, the accumulated miles driven has increased as witnessed by the increasing average age of vehicles on the road. Even though there has been a decline in total miles driven, the impact has

---

[5] *The Skipper's Course*, CG-433 (Washington, D.C.: U.S. Department of Transportation & United States Coast Guard, 1972), p. i.

[6] *Rate Training Manual*, NAVPERS 10085-B (Washington, D.C.: U.S. Government Printing Office, 1971), p. 28.

been least in around-town "stop & go" travel, which places greatest strain on automotive parts.

Furthermore, the trend in new car sales is increasingly toward the smaller, lighter vehicle which operates at higher RPM and engine temperature. These factors tend to cause parts to wear out faster, speeding up the replacement cycle and increasing demand.[7]

5. Examine the old-new information patterns in these two paragraphs. Do the sentences present old information first? From which part of the previous sentence do they draw the old information?

The vacuum advance mechanism has a spring-loaded diaphragm connected by linkage to the distributor breaker plate. The spring-loaded side of the diaphragm is airtight, and is usually connected to an opening in the carburetor. This opening is on the atmospheric side of the throttle plate when the throttle is in the idling position. In this position, there is no vacuum. [8]

• • •

Four of our well-equipped plants serving the hydrocarbon, petrochemical, and processing industries are strategically located in the southeast, midwest, and southwestern sections of the country. This means more economical and convenient shipping to any part of the country as well as points of shipment for export. The plants, while operating independently, coordinate their activities to provide the best products, services, and price to their customers. Each facility has the capability of producing fabricated steel products to virtually any design specification established by a customer, or we generate complete designs with our own engineering staff whenever special technology is required.[9]

---

[7]Arrow Automotive Industries *Annual Report 1980*, p. 63.

[8]*The Back-yard Mechanic*, Vol. II (Washington, D.C.: U.S. Government Printing Office, 1978), p. 55.

[9]Varlen Corporation *Annual Report: 1981*, p. 5.

Unit Eleven

Open Exercises

This unit consists solely of sentence-combining exercises. Not only will these exercises give you practice in creating stylistic prose, but they will familiarize you with a few of the forms of technical prose. This unit contains sample abstracts, factual summaries, conclusions, and descriptive reports, as well as letters of transmittal and application. Pay attention to the underlying organizational patterns in each of these samples so that you can employ similar strategies in your own writing.

Combine each exercise into a brief, full-length sample of technical prose. These are "open" exercises in that there are no rigid rules for making any of the combinations. Nor is there a correct way to combine the sentences. In deciding how to make each combination, keep in mind the probable *purpose* of and *audience* for the piece. Ask yourself: Is this combination the most appropriate way of expressing to the probable audience the particular message of this piece of writing? For example, if you are combining the letter of application on pages 165–167, in which the author seeks employment as a physics instructor, remember that your purpose is to argue for an interview and that your audience is the chairman of a university physics department and perhaps a committee of interviewers. Each sentence in the finished letter should specify coherently and articulately the credentials of the applicant in a tone that exudes confidence but not boastfulness.

Although there are no rules for combining these exercises, there are some factors you should be aware of. First, each group of sentences is prefaced by a multiple-decimal numbering system; the first numeral refers to the paragraph and the second to the sentence. For example, the number (3.4) informs you that the corresponding group of sentences constitutes the fourth sentence of the third paragraph. Nevertheless, you are not obligated to follow the suggested combining patterns. You may find, for example, that the sentences in groups (3.4) and (3.5) can be combined more effectively into one sentence than into two. Such considerations are entirely legitimate; these exercises are meant to stimulate your creativity, not to stifle it.

Second, you are not required to use the exact vocabulary of the exercise if you believe you have more effective alternatives. You may, for example, decide that it is more appropriate to use the verb *use* in place of *utilize*. As long as you remain faithful to the apparent meaning of the sample, you are free to manipulate the prose in any way you see fit.

In fact, if there is anything approaching a "rule" associated with these exercises, it is this: You are free to manipulate vocabulary, sentence structure, and organization in any manner *so long as your finished sentences contain every item of information represented by each sentence.*

Abstracts

Technical reports usually contain an abstract: a short description of the report that sometimes includes the report's main conclusions. The abstract appears on the front cover, the title page, or a separate page before the contents.

The following are three abstracts. Remember, the purpose of the abstract is to introduce a maximum amount of information in as short a space as possible; therefore, the language must be especially concise. Eliminate every irrelevant word.

(1.1) There is an increase in attendance.
The increase is substantial.
The attendance is at football games.
Seating is necessary.
The seating is additional.
The seating is at Salitsky Stadium.
The stadium is in San Bernardino, California.

(1.2) The two alternatives are additions.
The alternatives are considered in this study.
The additions are of seats.
The seats are 4,000.
Or the seats are 6,000.
The seats are in the end zone.
The end zone is in the north.
The end zone is in the stadium.

(1.3) This report discusses four criteria.
The criteria are main.
A criterion is parking facilities.
A criterion is costs.
The cost is of construction.

The cost is of maintenance.
The cost is of utility.

(1.4) The costs correspond to the number of seats added.
The costs are of construction.
The costs are of maintenance.
The costs are of utilities.
The corresponding is direct.

(1.5) The existing parking facility is adequate for the addition of 4,000 seats.
However, spaces are needed.
The spaces are for parking.
The spaces are additional.
The spaces are 700.
The spaces are for an addition.
The addition is of seats.
The seats are 6,000.

(1.6) The addition would cost less.
The addition is smaller.
The cost is in the short run.
But the addition would generate a return.
The addition is of seats.
The seats are 6,000.
The return is monetary.
The return is over a period.
The period is of years.
The years are ten.

(1.7) The study concludes the addition is cost-effective.
The addition is of seats.
The seats are 6,000.
The effectiveness is more.
It is more than an addition.
The addition is of seats.
The seats are 4,000.

• • •

(1.1) This study examines a heating system.
The system is solar.
The system is low-cost.
The system supplements a system.

The system is the library's.
The library is in the university.
The system is for heating.
The system is present.
The system is of warm-air.

(1.2) The heating system is completely automatic.
The system is solar.
The system consists of several components.
The components include.
They include piping.
They include a tank.
The tank is for storage.
They include a pump.
The pump is for water.
They include an exchanger.
The exchanger is for heat.
They include a collector.
The collector is solar.
The collector is flat-plate.

(1.3) The cost of the system is $80,000.
The cost is estimated.
The system is proposed.
It includes the cost.
The cost is of materials.
The cost is of installation.

(1.4) The materials needed for installation are available.
The installation is of a heating system.
The system is solar.
The availability is local.
The availability is at yards.
The yards are of lumber.
The availability is at hardware stores.
The availability is at supply stores.
The supplies are electrical.
The availability is at supply stores.
The supplies are for plumbing.

• • •

(1.1) This report presents results of research.
The research considers the purchase of wells.

The wells are No. 2, No. 4, and No. 6.
The wells are in Halli Field.
The field is West Kay County.
The county is in Pennsylvania.
The purchase is for $300,000.

(1.2) The report emphasizes shape and volume of the reservoir.
It emphasizes efficiency.
The efficiency is of recovery.
It emphasizes percentage.
The percentage is of decline.
The decline is in production.
It emphasizes taxes.
It emphasizes operating costs.
The costs are of the oil field.

(1.3) It calculates value.
The value is present.
The value is of each well.
And it includes value.
The value is of salvage.
The value is of the equipment.
The equipment is of the well.

Letters of Transmittal

The next three exercises are samples of a letter of transmittal: a short correspondence announcing either that a report is enclosed or that it is on its way under separate cover. Notice that in the typical transmittal letter, the introduction alerts the reader to the report's completion; the body briefly describes the report; and the conclusion invites the reader to contact the report's author if there are any questions.

Dear Ms. Dinan:

(1.1) You requested a report.
The request was on October 15, 1983.
I am submitting a research report.
The report is entitled "The Technical and Economic Advantages of Oil Agglomeration."

(2.1) This report compares two techniques.
   The techniques are for cleaning coal.
   The coal is fine.
   One technique is oil agglomeration.
   One technique is froth flotation.

(2.2) This report discusses percentage of energy recovery.
   The report discusses size and types of coal feeds.
   The report discusses degree and type of agitation.
   The report discusses pulp density.
   The report discusses coal-wetting properties.
   The report discusses production cost.

(2.3) The recommendation is based on these criteria.
   The recommendation is mine.

(2.4) My analysis shows a method.
   The method offers many advantages.
   The advantages are technical.
   The advantages are economical.
   Therefore, I propose installation.
   The installation is of a cleaning process.
   The cleaning is of coal.
   The coal is fine.
   The installation is by OLCO Inc.
   The installation is in OLCO's plant.
   The plant is for coal preparation.

(3.1) I am available for consultation.
   The consultation is about this report.
   The consultation is for further research.
   The consultation is while you consider techniques.
   The techniques are for cleaning.
   The techniques are alternative.

(3.2) I appreciate your consideration.
   The consideration is of this research report.
   I look forward to hearing from you.

(3.3) Call if you have any questions.
   Feel free to call me collect.
   Call at the number.
   The number is above.

• • •

Dear Mr. Ellingson:

(1.1) Enclosed is the report.
The report has a title.
The title is "Traffic Improvements at the Intersection of Fourth Avenue and Eighth Street."

(1.2) This report recommends traffic improvements.
The improvements should be implemented at an intersection.
The intersection is of Fourth Avenue and Eighth Street.

(2.1) The report analyzes the location.
The report analyzes criteria.
The report analyzes cost.
The cost is of restructuring the intersection.
The intersection is of Fourth Avenue and Eighth Street.

(2.2) The report has a primary concern.
The concern is to keep the cost within a budget.
The cost is of this project.
The budget is of the Dallas Traffic Commission.

(2.3) I have included a comparison of improvements.
The comparison is of cost.
The comparison is itemized.
The improvements are of traffic.
The improvements are the project's.
The improvements are proposed.

(2.4) The report also considers benefits.
The benefits are to the community.
The benefits will result from implementation of the project.

(3.1) I hope this report meets your approval.
The hope is sincere.

(3.2) I am a traffic engineer.
I am a motorist.
I travel through this intersection frequently.
I feel the volume is substantial.
It is substantial enough.
The volume is of traffic.
You should consider improvements.
The improvements are proposed.

(3.3) Any questions may arise.
　　　They may arise while you review this report.
　　　Please feel free to call.
　　　Please feel free to write.

(4.1) Thank you for consideration.
　　　The consideration is yours.

• • •

Dear Mr. Bottiger:

(1.1) Enclosed is the study.
　　　The study has a title.
　　　The title is "The Relative Values of Concrete and Asphalt for Surfacing the North Parking Lot of the Research and Development Building."

(1.2) I am submitting this report to help you choose.
　　　The choice is of a surface.
　　　The surface suits your needs.
　　　The suiting is best.

(2.1) This study provides an analysis.
　　　The analysis is complete.
　　　The analysis is of aspects.
　　　The aspects are negative.
　　　The aspects are positive.
　　　The aspects are of both surfacing agents.

(2.2) I have assumed.
　　　The assumption is in analyzing the cost.
　　　The cost is of surfacing.
　　　The surfacing is concrete.
　　　The assumption is about your company.
　　　Your company will perform the actual labor.

(2.3) However, I have assumed.
　　　The assumption is in analyzing the cost.
　　　The cost is of asphalt surfacing.
　　　The assumption is about a private contractor.
　　　The contractor will perform the labor.
　　　The labor is at a cost.
　　　The cost is specified per square yard.

(2.4) Asphalt surfacing must be contracted.
The contracting is because of equipment.
The equipment is specialized.
The equipment is required.

(2.5) I also include information in this study.
The information is on length of time.
The time is required for construction.

(3.1) I am certain.
This study will fulfill needs.
The needs are of your department.

(3.2) A problem may arise.
A question may arise.
Do not hesitate to contact me.
Contact me anytime.

Factual Summaries

Many technical reports contain a factual summary: a section outlining the relevant data of the report. Since the factual summary highlights specific items of information, its language is usually straightforward. The following are three sample summaries. Remember that readability is important in factual summaries, and combine these sentences accordingly.

(1.1) Locating the area was the first step.
The area is for landing.
The step is in the airport's design.

(1.2) The County Board has designated an area.
The county is New Haven.
The area is south of Meriden, Connecticut.
The area is the site of an airport.
The airport is proposed.

(1.3) The designated area is four miles south of the intersection.
The intersection is of interstates 40 and 13.

(1.4) The runway's length will be 3,500 feet.
The length is average.
The length is for a basic runway.
The runway is a utility-stage I.

(1.5) The runway will face northeast.
Northeast is a usual direction.
The direction is of prevailing winds.

(1.6) The runway's location depends on the height.
The height is of obstacles.
The obstacles are in the approach area.

(1.7) There must be a trajectory clearance.
The clearance must be 17 feet.
The trajectory is of take-off.
The clearance is over the interstate.
And the clearance is over public roads.
The clearance is 15 feet.

(1.8) The runway adheres to these criteria.
The runway is proposed.

(1.9) There are four reasons.
The reasons are principal.
The reasons are for locating the runway.
The location is in the proposed area.
(1) Noise will not disturb the areas.
The noise is of airplanes.
The areas are populated.
(2) The grade is consistent.
The grade is of the ground.
The consistency is fair.
(3) Clearing is necessary.
The clearing is little.
The clearing is of fields.
The fields are adjacent.
(4) The road needs to be extended.
The road is secondary.
The road is present.
The extension is only 1,500 feet.
The extension is to provide access.
The access is to the airport.

(1.10) The following sections will discuss all criteria.
The criteria are above.
The discussion is in detail.

• • •

(1.1) Friedlander Hall is the dorm.
   The dorm is for women.
   The dorm is the oldest.
   The dorm is at the University of Cincinnati.

(1.2) Fires have proven equipment faulty.
   The fires are past.
   The fires are at Friedlander.
   The equipment is for fire safety.

(1.3) This equipment includes smoke detectors.
   It includes fire extinguishers.
   It includes alarms.
   The alarms are to the station.
   The station is for fire.

(1.4) Friedlander Dorm is tall.
   It has 15 stories.
   It has a capacity of 900 occupants.

(1.5) Certain steps should be taken.
   The steps are in order to make a building fire-safe.
   The building is this size.

(1.6) First, fire equipment must be removed.
   The equipment is old.
   The equipment is faulty.

(1.7) Second, equipment must be purchased.
   Equipment must be installed.
   The equipment is new.

(1.8) Third, a building inspector must be hired.
   The inspector is monthly.
   The inspector inspects all safety equipment.
   The equipment is for fires.

(2.1) Contracting would cost $3,200 to $4,100.
   The contracting is of an inspector.
   The inspector is of buildings.
   The cost is rough.
   The cost is per year.

(2.2) An alternative is to hire a maintenance superintendent.
The superintendent is from the University of Cincinnati.
The hiring is for this job.

(2.3) The annual stipend is $900.
The stipend is for a maintenance superintendent.

(3.1) We should consider purchasing smoke detectors.
We should purchase fire extinguishers.
The purchase is from Ohio Protective Services.

(3.2) Ohio Protective Services will discount the price of equipment.
The equipment is new.
The discount is in return.
The return is of equipment.
The equipment is old.
The equipment is faulty.
The equipment is in Friedlander.

(3.3) The cost is approximately $8,500.
The cost is with a discount included.
The cost is for safety equipment.
The equipment is for fires.
The equipment is new.

(3.4) A payment plan is available.
The plan is long-term.
The plan is available through OPS.

(3.5) This project is not a question of convenience.
It is one of safety.

• • •

(1.1) Woodlawn Cafe's kitchen was built according to specifications.
The specifications were the Health Department's.
The specifications were of 1972.
The kitchen does not meet present requirements.
The requirements are in areas.
The areas are of ventilation.
The areas are of surface.
The surface is of floors.
The areas are of ratio.
The ratio is of restrooms.
The ratio is to employees.

(2.1) The ventilation system should be updated.
The system consists of the stove's hood.
It consists of the fan.
The fan is for ventilation.
The ventilation is exterior.

(2.2) The stove's hood does not meet the requirement.
The hood is present.
The requirement is the Health Department's.
The requirement is of drawing 500 cubic centimeters.
The centimeters are of air.
The drawing is every 10 seconds.

(2.3) The installation is to correct this deficiency.
A hood should be installed.
The hood is a Hartsall XL 2001.
The hood is the stove's.

(2.4) The stove's hood will cost $1,065.
The cost is plus a shipping charge.
The charge is $30.
The shipping is from Atlanta, Georgia.

(2.5) It is according to the Health Department.
A fan must be installed.
The fan is for ventilation.
The ventilation is exterior.
The fan is to draw heat.
The fan is to draw odors.
The drawing is from the kitchen.

(2.6) A fan will cost $531.
The fan is for ventilation.
The ventilation is exterior.
The fan is a Hartsall SD 95.

(3.1) The present floor is a potential hazard.
The floor is concrete.
The floor is in the kitchen.
The hazard is to employees.
The hazard is because of a possibility.
The possibility is of someone slipping.
The possibility is of someone injuring himself.
The injuring is on the hard concrete.

(3.2) The installation eliminates this problem.
A floor should be installed.
The floor is tile.
The tile is quarry.
The floor is fabricated.
The fabrication is special.

(3.3) This tile is designed to produce traction.
The traction is better than concrete.
It is better even when wet.

(3.4) Also, the tile will channel water.
The channeling is to drains.
The drains are in the floor.

(3.5) We can use Dap cement.
The cement is for floors.
The cement prevents leaking between tiles.

(4.1) It is according to the Health Department.
Another restroom must be built.
The restroom is for employees.

(4.2) It should be located.
The location is between the office.
The location is between the present restroom.

(4.3) The construction cost will be minimal.
It is minimal for two reasons:
(1) Only one wall must be built.
The wall is a partition.
(2) Sewage pipes can be tied into a line.
The line is for sewage.
The line is the adjacent restroom's.

(5.1) We should add drains.
The drains are in the floor.
The drains are eight-inch.

(5.2) The Health Department requires one drain.
The drain is in the new bathroom.
Two are needed in the kitchen.

(6.1) These changes will meet specifications.
The meeting is not only.

The specifications are of 1984.
The specifications are the Health Department's.
They will increase efficiency.
The efficiency is total.
The efficiency is of the cafeteria.

Letters of Application

The following are two letters of application for employment. One is for a position as a physics instructor; the other is for employment as an engineer at a North Sea oil operation. It is important in application letters to use an appropriate tone, one that is confident but not cocky, respectful but not overly so. Keep in mind the importance of tone when you combine these sentences. In addition, you may wish to pay special attention to the clear and logical organization of these letters.

Dear Mr. Wills:

(1.1) Dr. Gary Buck has informed me.
 Buck is the Chairman of the Physics Department.
 The department is at Tech University.
 You are seeking an individual.
 The individual fills a faculty position.
 The position is as a Laboratory Instructor.

(1.2) I wish to apply.
 The application is for this position.

(1.3) I am confident.
 I have experience.
 I have background as well.
 The background is theoretical.
 They are to fill the position.
 The filling is competent.
 The filling is effective.

(2.1) My experience includes work as a Graduate Assistant.
 The work was at Tech University.

(2.2) This position involved laboratory research.
 It involved tutoring students.
 The tutoring was in physics.
 It involved grading.

　　　　　The grading was of homework and laboratory reports.
　　　　　It involved proctoring examinations.

(2.3)　I gained experience as a Graduate Assistant.
　　　　　The experience is valuable.
　　　　　The experience is in helping students.
　　　　　The students understand aspects.
　　　　　The aspects are difficult.
　　　　　The aspects are many.
　　　　　The aspects are of physics.

(3.1)　A major part is in the fields.
　　　　　The part is of my professional concentration.
　　　　　The fields are of physics.
　　　　　The physics is atomic.
　　　　　The physics is nuclear.

(3.2)　My thesis involved investigation and utilization.
　　　　　The investigation and utilization were of reactions.
　　　　　The reactions were (P,Y).
　　　　　The reactions were various.
　　　　　The reactions were a means of calibrating.
　　　　　The calibration was of an accelerator.
　　　　　The accelerator was Tech's.
　　　　　The accelerator was a 2 MeV Van de Graaff.

(3.3)　I have studied Quantum I, II, and III.
　　　　　I have studied physics as well.
　　　　　The physics is nuclear.
　　　　　I have studied mechanics.
　　　　　The mechanics is statistical.

(3.4)　Also, I have studied electricity.
　　　　　I have studied magnetism.
　　　　　I have studied physics.
　　　　　The physics is theoretical.
　　　　　I have studied mechanics.
　　　　　The mechanics is advanced.
　　　　　I have studied physics.
　　　　　The physics is solid-state.
　　　　　The studying was as a graduate student.

(3.5)　As an undergraduate, I was involved.
　　　　　The involvement was at Bloomsburg State College.

The involvement was active.
The involvement was in experimenting.
The experimenting was in utilization of solar energy.
The energy was for heating.
The heating was of water.
The heating was domestic.
The water was hot.

(3.6) Part of my study concerned measuring the efficiency.
The efficiency is of a heater.
The heater is of water.
The heater is solar.
I designed the heater.
I constructed the heater.

(4.1) I feel.
My experience equips me.
The experience is as a Graduate Assistant.
The experience is coupled with my background.
The background is academic.
They equip me to be an instructor.
The instructor is of physics.
The instructor is effective.

(5.1) Enclosed is a résumé.
The résumé is mine.

(5.2) Call if you desire additional information.
Call if you wish to schedule an interview.
Feel free.
Call collect.
Call at the above number.

(6.1) Thank you for consideration.
The consideration is yours.

• • •

Dear Mr. Jones:

(1.1) Your advertisement invited applications.
The advertisement is in an issue of *Oil and Gas Journal*.
The issue is January 1984.
The applications are for the position of off-shore drilling engineer.

The applications are for fluid-control engineer.
The engineer is for operations.
The operations are in the North Sea.
The operations are yours.

(1.2) I would like to apply.
The application is for this position.

(2.1) In August 1984, I will receive a BS degree.
The degree is in petroleum engineering.
The degree is from the University of Georgia.

(2.2) My area of specialization is in techniques.
The techniques are of off-shore drilling.
It includes well-logging.
The well-logging is high-salinity.
It includes string-targeting.
The string-targeting is multidrill.
It includes pressure interpretation.
The pressure interpretation is deep-well.

(2.3) I have worked for a year with an engineering firm.
The firm is in Virginia.
The firm is Engineering Associates.
The work is in the field of fluid-flow control.

(2.4) My job included flow monitoring.
The monitoring is of a sewer system.
The system is of a city.
It included chemical analysis as well.
The analysis is of material.
The analysis is at various flow rates.

(3.1) I am interested in the position of drilling engineer.
However, I noticed something in your advertisement.
You are also seeking personnel.
The personnel are for the financial end.
The end is of your North Sea Operation.

(3.2) I received my degree in 1979.
The degree is in international economics.
The degree is from the University of Georgia.

(3.3) It was in the curriculum of economics.
I studied markets.

The markets are capital.
I studied management.
The management is financial.
The management is international.

(3.4) It is with this additional business background.
I may be of service to you.
The service is in several aspects.
The aspects are of your North Sea production.

(4.1) The advertisement places strong emphasis.
The emphasis is on willingness to relocate to Great Britain.

(4.2) It was in 1980.
I spent a full term at the University of Sussex.
Sussex University is in Brighton, England.

(4.3) It was while attending Sussex.
I studied markets.
The markets are banking.
The banking is British.
The markets are financial.

(4.4) I would enjoy returning.
The enjoyment is great.
The returning is to England.
The returning is to work for you.

(5.1) Engineering Service Company is an organization.
The organization is fine.
I would like to be a part of it.

(5.2) It is if the position is still available.
Please contact me.
I will be happy to come to New York.
I will come at your convenience.
I will come for an interview.

(5.3) My address is printed.
My telephone number is printed.
The printing is on a résumé.
The résumé is enclosed.

(5.4) I am looking.
The looking is forward to hearing from you.

Conclusions

Most technical reports contain a conclusion section, which can take the form of prose paragraphs, a numbered list of concluding statements, or a mixture of both formats. The following are three sample conclusion sections. The first two are prose, and the third is a mixture of prose and lists. The third exercise is particularly difficult because it deals with many numbers and numerical relationships. Examine the sentences carefully to determine exactly where each of the modifiers belongs.

(1.1) Our analysis shows.
 The computer can perform functions.
 The computer is an Apple II Plus.
 The functions are needed.
 The need is to replace our system.
 The system is time-sharing.
 The replacement is at a substantial savings.

(1.2) It can be used to write programs.
 It can be used to process preprogrammed packages.
 They are for handling all functions.
 The functions are of our business.

(1.3) The system offers enough memory.
 The system is an Apple II Plus.
 The memory accommodates all our needs.
 It can be expanded as the company grows.
 The company requires more memory storage.

(1.4) Also, the Controller Business Package offers accounting programs.
 The programs are a wide variety.
 The programs handle needs.
 The needs are of our business.

(1.5) With the addition, the department can become self-sufficient.
 The addition is of the Apple Writer.
 And the addition is of the suggested printer.
 The department is of accounting.
 The self-sufficiency is from a time-sharing system.

(1.6) The cost is $9,033.
 The cost is of purchasing a system.
 The system is complete.

(1.7) A warranty will cost $53.
    The warranty is extensive.
    The cost is per month.

• • •

(1.1) There are many benefits for our company.
    The benefits are if we were to apply a method.
    The method is Direct-Digital Logging.
    The method is applied to our production.

(1.2) We would not have to worry about overriding.
    Inflation overrides profits.

(1.3) The years are after operation begins.
    The years are ten.
    We will average approximately $6 million.
    The dollars are per producing well.

(1.4) There will be fewer injuries on the job.
    The injuries are due to various safety systems.
    The systems are discussed in this report.

(1.5) Time will be reduced.
    The time is spent at each oil site.
    The reduction is great.
    The reduction saves money for the company.

(1.6) It is with profit.
    The profit is large.
    The profit is expected from this method.
    We can expand production.
    We can create opportunities.
    The opportunities are for additional employment.
    The opportunities are in the community.

(1.7) It is if the method is given a chance.
    The method is Direct-Digital Logging.
    It is if it is given the allotted amount of time.
    It will be a great boon to the company.

• • •

(1.1) There has been a significant number of accidents.
    The accidents are at the interchange.
    The interchange is of Interstate 84 and Route 8.

(1.2) Police accident reports show.
　　　The police are of Waterbury.
　　　Accidents were due to a combination of defects and conditions.
　　　The accidents are 1,932.
　　　The accidents were last year.
　　　The defects were of highway design.
　　　The conditions were of a road.

(1.3) Spot checks reveal frequent traffic congestion.
　　　The checks are of the interchange.
　　　The congestion is at the exits and entrances.

(1.4) It is in addition to design defects.
　　　The interchange does not conform to standards.
　　　The standards are of the Federal Highway Association (FHWA).
　　　The standards are of the Connecticut Highway Department (CHWD).

(1.5) The FHWA clear-roadside concept requires.
　　　The requirement is for all objects.
　　　The objects are fixed within 30 feet.
　　　The feet are within the traveled way.
　　　They are removed or made "breakaway."

(1.6) Otherwise, they are to be protected.
　　　The protection is by guardrail.
　　　The protection is by impact-attentuation devices.

(1.7) There are four design aspects of the interchange.
　　　The aspects do not meet this requirement:
　　　(1) The median currently has no end treatment.
　　　The median is of a ramp.
　　　(2) The bridge abutment is located.
　　　The location is 12 feet from the traveled way.
　　　(3) The present lighting posts are located.
　　　The location is 20 feet from the traveled way.
　　　They do not have a "breakaway" design.
　　　(4) The exit signs are located at the edge.
　　　The edge is of the recovery areas.
　　　They do not have a "breakaway" design.

(1.8) There are seven design aspects.
　　　The aspects do not meet CHWD requirements:
　　　(1) The deceleration lanes are long.

The lanes are at Exit 24.
The exit is from Interstate 84.
The length is 300 feet.
The specifications require a length of 430 feet.
(2) The deceleration lanes are long.
The lanes are at Exit 5.
The exit is from Route 8.
The length is 175 feet.
The specifications require a length of 265 feet.
(3) The acceleration lanes are long.
The lanes are on to Interstate 84.
The length is 550 feet.
The specifications require a length of 400 feet.
(4) The recovery areas are 8 feet wide.
The areas are at the Interstate 84 exit.
The exit tapers.
The tapering is from 120 feet.
The tapering is to the running lanes.
(5) The recovery areas are nonexistent.
The areas are at the Route 8 exit.
The nonexistence is virtual.
(6) The shoulder is 5 feet wide.
The shoulder is between Interstate 84's exit and entrance.
The standards require a shoulder width of 10 feet.
(7) There are no shoulders along the ramps.
The ramps are for exit and entrance.
The standards require a shoulder width of 5 feet.

(2.1) I conclude.
   A renovation would improve traffic.
   The renovation is of an interchange.
   The interchange is of Interstate 84 and Route 8.
   The renovation would decrease the number of accidents.
   The decrease is significant.
   The accidents are at the interchange.
   The renovation would bring the interchange within standards.
   The standards are present.

(2.2) The interchange qualifies for aid.
   The interchange is of Interstate 84 and Route 8.
   The aid is from a Fund.
   The fund is for Improvement.
   The improvement is of Highways.
   The highways are in Connecticut (FICH).

Descriptive Reports

The following is a short descriptive report about a certain type of flashlight (see Figure 1). The report adheres to a panel format; that is, each section is clearly delineated by a topic heading. Notice that the paper follows a traditional organizational pattern for descriptive reports: It begins with a definition, proceeds to separate sections detailing the main parts of the subject, and ends with a discussion of how the subject operates. Notice, too, that the report contains many numbers and measurements. Incorporating them into your prose so that the report reads smoothly is part of the challenge of this exercise.

The GA-40 Flashlight

Introduction

(1.1) A flashlight is a portable light source.
The light is electric.
It makes vision possible in the absence of other light sources.

(1.2) The light is in the form of a stream of radiation.
The radiation is electromagnetic.
The radiation is reflective.

(1.3) The greatest asset is its portability.
The asset is the flashlight's.

(1.4) Most varieties are designed to be handheld.
They are small compared to other sources of light.
The sources are such as table lamps.

(1.5) Another virtue of the flashlight is its energy source.
The source is self-contained.
The source is batteries.
The batteries are standard size.

(1.6) There are several types of flashlights.
They vary in size.
They vary in shape.
They vary in number of cells.
They vary in size of cells.
The cells are of energy used.
But the most common version operates on two D-size cells.
The version is such as the GA-40.

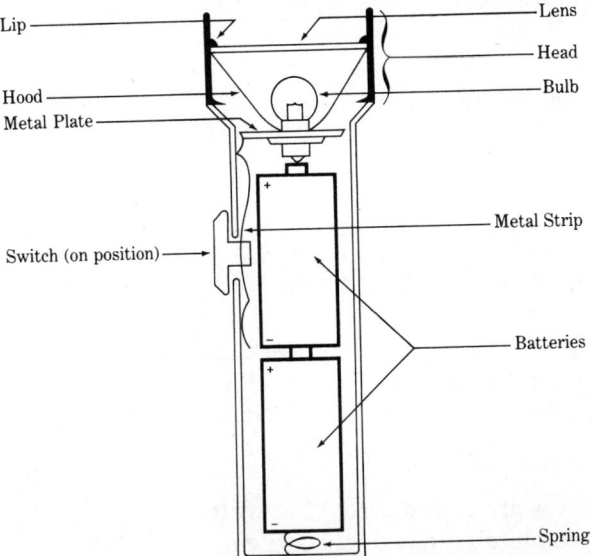

Figure 1 GA-40 Flashlight (Cross-Section View)

(1.7) The GA-40 has four basic parts.
  A part is the body.
  A part is the switch.
  A part is the bulb housing.
  A part is the energy source.

Body

(2.1) The body of the GA-40 is a tube.
  The tube is metal.
  The tube is open at one end.
  The tube is closed at the other.

(2.2) The tube has a length.
  The length is 16.5 centimeters.
  The tube has a diameter.
  The diameter is 3.9 centimeters.
  (See illustration.)

(2.3) The open end flares out.
  The flaring is to a diameter.
  The diameter is 4.3 centimeters.
  It is threaded on the outside.
  The threading is so that the bulb housing can screw onto it.

(2.4) Attached is a spring.
    The attaching is to the inside.
    The attaching is to the bottom of the body tube.
    The spring is steel.
    The spring is slightly smaller than the tube itself.
    The smallness is in diameter.

Switch

(3.1) There is a switch behind the threaded end.
    It is 4 centimeters behind.

(3.2) This plastic rectangle slides toward the threaded end.
    The sliding activates the light.
    The sliding is back toward the closed end.
    The sliding extinguishes the light.

(3.3) Tiny grooves are etched across the switch.
    It is so the thumb can slide the switch.
    The sliding is firm.

(3.4) Underneath the switch is a hole.
    The hole is rectangular.
    The hole is small.
    The hole is in the body of the flashlight.

(3.5) A small bar extends from the switch through this hole.
    It attaches to a strip.
    The strip is thin.
    The strip is copper.
    The strip is inside the tube.

(3.6) This strip runs parallel to the body.
    It serves to hold the switch in place.
    It completes the electric circuit.
    The circuit operates the bulb.

Bulb Housing

(4.1) The bulb housing consists of a plastic cylinder.
    The housing consists of a plastic lens.
    The lens is clear.
    The housing consists of a reflective cone.
    The housing consists of a bulb.

(4.2) The plastic cylinder is 5 centimeters in diameter.
    The cylinder screws onto the body of the flashlight.

(4.3) Halfway into the cylinder is a lip.
The lip is .2 centimeters.
The lip supports a lens.
The lens is plastic.
The lens is clear.

(4.4) The lens holds the components of the flashlight.
The holding is in the tube.
The holding is when the housing is screwed onto the body.

(4.5) A cone rests against the inside of the lens.
The cone is coated with a highly reflective substance.
The substance diffuses the bulb's light.

(4.6) This cone contains a hole.
The hole is 1 centimeter in diameter.
The hole is rather than a tip.
The hole is so that a small bulb can protrude.
The protruding is into the interior.
The interior is reflective.
The interior is the cone's.

(4.7) The opposite end of the bulb rests against one of the batteries.
It is held by a plate.
The plate is metal.
The plate is circular.

Energy Source

(5.1) The batteries are called D-cells.
The batteries are in the GA-40.
It is a designation of their size.

(5.2) Each battery produces 1.5 volts.
The production is when fully charged.
It provides the bulb with electric potential.
The potential is of 3 volts.

(5.3) The first battery stands inside the body tube.
The battery's end is against the spring.
The end is flat (negative).
The spring is fixed to the tube's bottom.

(5.4) The other battery stands in the tube.
Its flat end is on top of the first battery.
Its positive end pushes against the bulb.

(5.5) The tube's spring pushes the batteries.
   It pushes snugly.
   They are against one another.
   They are against the bulb.

Operation

(6.1) Operation of the flashlight involves completion of a circuit.
   The circuit is electric.
   The circuit is simple.

(6.2) Push the switch.
   The pushing is toward the bulb housing.

(6.3) This slides the metal strip against the plate.
   The strip is inside the tube.
   The plate is metal.
   The plate is circular.
   The plate is of the bulb.

(6.4) An electric charge travels through both batteries.
   It travels into one electrode of the bulb.
   It travels through the filament.
   It travels out the other electrode.

(6.5) The spent charge then travels from this electrode.
   It travels through the metal cylinder of the bulb.
   It travels across the metal plate.
   The plate is circular.
   It travels into one end of the metal strip.
   The strip is the flashlight's.

(6.6) The charge then flows through the strip.
   It flows down into the spring.
   The spring transmits it back into the negative end.
   The end is of the first battery.

(6.7) The electric charge causes electrons to react.
   The charge is passing through the thin filament.
   The filament is of the bulb.
   The reacting is with one another.
   They emit light.
   The light is in the form of radiation.
   The radiation is electromagnetic.

(6.8) This light is diffused.
  The cone diffuses.
  The cone is reflective.
  The diffusing produces a beam.
  The beam is of light.

(6.9) It terminates operation.
  The operation is of the GA-40.
  Pull the switch back.
  It is pulled to its original position.

• • •

The following is a brief descriptive report about a common engineering tool: the drafting compass (see Figure 2). Combine the sentences while trying to retain the accuracy of the description. Take special care that the modifiers in your final report modify what they originally were meant to.

Introduction to the Drafting Compass

I. Definition and General Description

(1.1) A drafting compass is an instrument.
  The instrument is used for drawing circles.
  It is used for transferring measurements.
  The transferring is to paper.

(1.2) The tool consists of three basic parts.
  A part is a leg.
  The leg is for anchoring.
  A part is a leg.
  The leg is for drawing.
  A part is a pivot (see illustration).

(2.1) Each of the two legs has a point at one end.
  The point is sharp.
  Both are connected by a pivot.
  The pivot is at the opposite end.

(2.2) The parts are made of steel.
  They are coated with enamel.
  The enamel is corrosion resistant.

(2.3) This enamel is smooth.
  It is shiny.

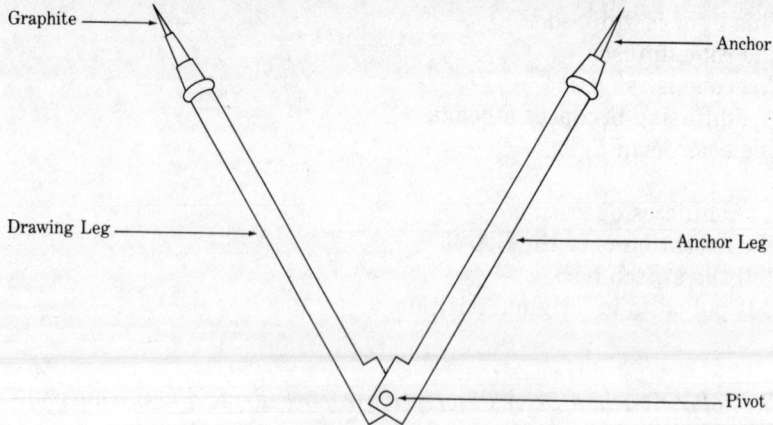

Figure 2   Drafting Compass

It is silver.
The silver is metallic in color.

(2.4) It is with the ends of the legs pointed upward.
The ends are pointed.
The ends are sharp.
The legs are separated an inch or two.
The compass resembles the letter "V."
It is with a toothpick.
The toothpick is steel.
The toothpick is at one tip.
It is with a piece.
The piece is of lead pencil.
The piece is at the other.

II. Anchor Leg

(3.1) The anchoring leg is a tube.
The tube is steel.
The tube is hollow.
The tube has a diameter about that of a pencil.
It is about 3½ inches in length.

(3.2) A cylinder extends from the leg.
The cylinder is steel.
The cylinder is ¼ inch long.

(3.3) Protruding from this cylinder is a point.
The point is steel.

The point resembles a toothpick.
The toothpick is round.
The toothpick is standard.

(3.4) This toothpick end is used as a reference point.
The end is sharp.
It is for taking a measurement.
Or it is to anchor the compass.
The anchoring is on a piece of paper.

III. Drawing Leg

(4.1) The drawing leg is identical in dimensions to the anchoring leg.
It has a similar cylinder.
The cylinder is steel.
The cylinder is fitted into one of its ends.

(4.2) This cylinder contains a cylinder.
The cylinder is plastic.
The cylinder is tapered.
The cylinder grips a piece of graphite.
The grip is tight.
The graphite is sharpened.
The graphite is the size of small pencil lead.

(4.3) The overall length is 3 inches.
The length includes fittings in the end.
The length is shorter than the anchoring leg.
The shortness is about ½ inch.

(4.4) This difference allows the draftsman to adjust the amount of graphite.
The difference is in length.
The graphite protrudes from the leg.

IV. Pivot

(5.1) The ends of the legs are flattened.
The ends are opposite the tips.
The tips are sharpened.
The tips are theirs.
The ends form a vertex.

(5.2) A small rivet attaches the two ends.
The ends are flattened.
The rivet serves as a pivot for both legs.

(5.3) The distance of the tips determines size.
   The tips are pivoted away from one another.
   The size is of the measurement being transferred.
   Or it determines the radius.
   The radius is of the circle being drawn.

(5.4) This distance ranges in inches.
   The range is from ½ to 5½.

V. Functions

(6.1) It is when drawing circles with a compass.
   Place the tip on the desired centerpoint.
   The tip is of the anchoring leg.
   Place the tip against the surface to be drawn on.
   The tip is graphite.
   The tip is of the drawing leg.

(6.2) Grasp the pivot by the thumb and forefinger.
   Rotate the compass until you draw a circle.
   The circle is full or partial.

(6.3) It is to transfer a measurement.
   Pull the two compass tips apart.
   Rest one tip on one end of the figure.
   The figure is to be measured.
   Rest the other tip at the opposite end of the figure.

(6.4) The space represents the measurement.
   The space is between the two ends.
   The ends are of the compass.
   The measurement is of the figure.

(6.5) You now place the compass on the surface.
   The measurement is being transferred.
   The transferring is to the surface.

(6.6) The graphite will leave a tiny dot.
   The other tip can be used to punch a tiny hole.

(6.7) The distance between the hole and dot is the length.
   The length is the same as the original measurement.

Glossary

active voice When the subject of a sentence does the action the verb describes; for example, "Joann *kissed* Jerry." (Also see *passive voice*.)

addition/deletion The process of adding some information of one sentence to another while deleting the unnecessary words from both.

average sentence length (ASL) The average number of words per sentence in a piece of writing or publication. The ASL can have a direct effect on readability.

binder A word or phrase that helps connect sentences in a paragraph cohesively.

clause A group of words containing a subject and a predicate.

coordinate strategy The process of joining two or more words or sentences by using key coordinate words, such as *and, but, or*, and so on.

degree of variance (DOV) The number of words in the shortest and longest sentences in a piece of writing when one is trying to determine the average sentence length.

dummy subject The word *it* or *there* (usually used along with a form of the verb *to be*) acting as a replacement for the subject of a sentence; for example, "*It* is clear that...." and "*There* are two bolts fastened...."

embedding The process of inserting the facts or bits of information of one sentence into another sentence.

ing/ed-connection A verb form ending in *ing* or *ed* that usually functions as an adjective.

left branching The process of attaching words to the left part of a sentence, often with a comma separating the sections.

location marker A binder that points to a location in space (*here, there,* and so on) or time (*now, then,* and so forth).

middle branching The insertion of words into the middle of a sentence; words are often set off with two commas.

nominalization A verb or adjective that has been changed into a noun; for example, *puzzlement, oxidation, reduction.* Nominalizations often decrease readability.

noun cluster Several nouns grouped together without the usual connective words between them. For example, "the chemical environment protection agency." Noun clusters can decrease readability.

noun/noun connection A noun (and the words attaching to it) that renames another noun in a sentence.

paragraph cohesion The degree of unity in a paragraph, that is, how well (or poorly) the sentences are arranged in relation to one another in order to achieve an effective paragraph.

passive voice When the subject of a sentence is acted upon instead of performing the action the verb describes; for example, "Jerry *was kissed* by Joann." (Also see *active voice.*)

personal pronoun A word that takes the place of a noun and that refers to a person; for example, *I, you, he, she, we,* and so on.

phrase A group of words without a subject or predicate, usually used as a single part of speech.

prepositional phrase A preposition (*of, to, from, in, out,* and so forth) joined to a noun or a pronoun and functioning as a single word in a sentence; for example, "in the machine," "from the boss," "on the table."

preposition/deletion strategy The process of combining two sentences while using a preposition to change one of the sentences to a phrase.

readability The ease or difficulty with which a sentence or passage is read. For example, a long sentence about a complex subject that also contains difficult or unfamiliar words may have lower readability than a shorter, simpler sentence.

redundancy Unnecessary repetition of words or ideas. For example, "past history" is redundant because the same idea is expressed in both words; one of the terms would suffice.

repetition Reusing key words occasionally in a paragraph or a sentence to help ensure coherence.

right branching The process of attaching words to the right half of a sentence, often with a comma separating the sections.

signpost binders A word or phrase that serves as a transition within or between sentences (for example, *however, on the other hand,* and so on).

stylistic options The choices of wording and structure a writer has when writing.

subordinate clause A clause that cannot stand by itself as a complete sentence.

subordinate strategy An operation in which a sentence is added to the main sentence so that the added sentence is clearly dependent on the main one. The main sentence establishes the primary meaning in the combined sentence, and the added sentence qualifies or modifies the meaning of the main sentence.

substitute A word, often a pronoun, that stands for another noun in a sentence.

transforming The process of changing a word or grammatical relationship from one form to another; often used when embedding one sentence into another.

transposing The process of taking a word, phrase, or clause from one part of a sentence and moving it to another part of the same sentence.

variation An alternative form of a word used in a single paragraph to help ensure cohesion.

wh-connection A group of words that, like an adjective, modifies either a noun or a pronoun. The five connecting words introducing the wh-connection are *which, who, whom, whose,* and *that.*

About the Authors

James DeGeorge serves on the journalism faculty at Indiana University of Pennsylvania, teaching technical and industrial writing. He has developed for the department a number of courses in document design, research techniques for industrial writers, and presentation-making skills. Prior to joining the journalism faculty, he was Director of Graduate Studies in English and helped develop the department's Ph.D. program in rhetoric. He has authored numerous articles on writing-related issues and is also a consultant in communication matters to business and industry.

Gary A. Olson teaches courses in technical writing and in rhetoric and composition at the University of North Carolina at Wilmington, where he also directs the Center for Writing. He has published widely on writing-related issues in such periodicals as *College English, Journal of Advanced Composition, Teaching English in the Two-Year College*, and *CEA Critic*. He frequently serves as a technical-writing consultant, and he is an associate editor of *Technical Communication*.

Richard Ray is an English professor at Indiana University of Pennsylvania, where he teaches technical writing and does applied research for The Graduate School. For several years now he has also been conducting research in modern rhetoric, particularly as it applies to readability in technical and business writing. He has been a writing teacher for more than 20 years and has done industrial consulting in both written and oral communication.